Handbook of Accurate Surveying Methods

S P COLLINS
BSc Tech, AMCT, CEng, FICE

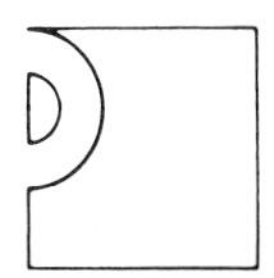

Pitman Publishing

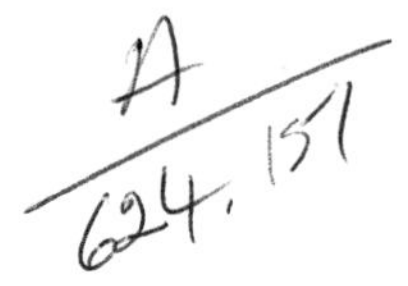

First published 1972

SIR ISAAC PITMAN AND SONS LTD.
Pitman House, Parker Street, Kingsway, London, WC2B 5PB
P.O.Box 46038, Portal Street, Nairobi, Kenya

SIR ISAAC PITMAN (AUST.) PTY. LTD.
Pitman House, 158 Bouverie Street, Carlton, Victoria 3053, Australia

PITMAN PUBLISHING COMPANY S.A. LTD.
P.O.Box 11231, Johannesburg, S. Africa

PITMAN PUBLISHING CORPORATION
6 East 43rd Street, New York, N.Y.10017, U.S.A.

SIR ISAAC PITMAN (CANADA) LTD.
495 Wellington Street West, Toronto, 135, Canada

THE COPP CLARK PUBLISHING COMPANY
517 Wellington Street West, Toronto, 135, Canada.

ISBN: 0 273 25214 3

G2-(T.1269:73)

PRINTED BY UNWIN BROTHERS LIMITED, OLD WOKING, SURREY

Preface

Engineers and technicians are frequently entrusted with the mapping of areas where engineering projects may be developed. They are concerned also with the creation of survey frameworks to control these projects during construction or to record their behaviour after completion. The increasing scale and complexity of modern schemes demands a sureness, precision and speed where failure in any one of these aspects is expensive and time-consuming.

The systems described here were developed by the author on a wide variety of engineering works including tunnels, bridges, dams and harbours.

Their success springs from: careful reconnaissance for well-proportioned, closed networks, sited to avoid distortion by atmospheric and other field effects; the establishment beforehand of standard methods of observing, recording and computing to make these functions quasi-automatic; design of those methods so that they will monitor the accumulation of errors and will be self-checking to avoid mistakes; the removal of ambiguity whereby bearings are unique and all operations and corrections for error are applied systematically in the correct sign.

The purpose of this book, then, is to extend the theoretical knowledge of the reader to cover the practical work which he will meet in the field. Descriptions of standard surveying instruments and their adjustment, for example, are not included where they are adequately treated in other books. The examples are given in SI (metric) units but may be used without any modification under the Imperial system; constants and standards are given under both headings in the appendices. The conversions are not exact in all cases as some of the figures are adjusted suitably in order to conform with standard practice.

Contents

Preface
1 THE STEEL BAND · 1
2 THE THEODOLITE · 5
3 ELECTROMAGNETIC DISTANCE MEASUREMENT · 11
4 LEVELLING · 15
5 TRAVERSE SURVEYS · 22
6 TRIANGULATION SURVEYS · 33
7 GEODIMETER SURVEYS · 45
8 THE THREE POINT PROBLEM · 58
9 TACHEOMETRIC SURVEYS · 62
10 SHAFT PLUMBING · 65

APPENDIXES

I Bibliography · 78
II Constants · 78
III Curvature and refraction corrections for vertical angles · 79
IV Length/altitude corrections · 80
V Proportional parts for seconds of arc · 81
VI Tacheometric survey table · 82
VII Crandall's method for adjusting a traverse · 84
VIII Notes on the use of the Geodimeter · 85
IX Geodimeter - measuring example · 90
X Geodimeter calibration tables · 91
XI Geodimeter - Nomogram for atmospheric corrections · 96
XII The Kern ME 3000 Mekometer · 97
XIII Geodimeter field sheet · 98

1 The Steel Band

A variety of steel bands or measuring tapes is available and, whilst good results may be obtained with 60 or 100 m bands, 13 mm wide, they appear to be much more susceptible to error except in the hands of very experienced observers. It is highly unlikely that accurate measurements will be obtained using steel band chains 20 m long and 16 mm wide.

Except for short lengths, accurate results are obtained most conveniently with a steel band 100 m (300 ft) long, 7 mm (1/8 in) wide, marked at every centimetre and provided with handles outwith the zero marks. The method set out below is based mainly on the use of this type.

In conjunction with the band a spring balance calibrated up to 10 kg (30 lb), a thermometer and a roller grip are required.

The accuracy of measurements using a band made of steel is governed mainly by the accuracy with which the temperature of the band is known. Field experiments carried out at the National Physical Laboratory and elsewhere show that the band temperature will differ quite unpredictably from the thermometer temperature by about 3°C on a calm, dry, overcast night and by up to 15°C on a hot sunny day. Thus the accuracy of a measurement cannot be guaranteed to be of higher order than 1/30000 and may be as low as 1/6000 through temperature errors alone. It is interesting to note also that the Ordnance Survey concede an overall accuracy no higher than 1/20000 for this work.

All precise work therefore should be carried out at night, when conditions are better and also steadier than by day.

Further sources of error arise in the laying-out of the band. It should lie straight between the end points or be made to lie straight between points of horizontal or vertical deviation within the length of the band. These deviations must be measured accurately. A thorough inspection along the length of the band is made to eradicate (or measure) horizontal kinks and to select the points whose offset is to be measured for lateral deviation or levelled for slope correction. Undisclosed deviations within the length of the band give cumulative and positive errors and may be large. For example, misalignment of 100 mm at the mid point of a 5 m length requires a correction of 1/25000 when using a 100 m band, 1/15000 for a 60 m band.

Where necessary the band should be supported at intervals of about 5 m to avoid the need for catenary corrections which are too common a source of mistakes. Instead then a simple, arbitrary sag/irregularities correction is applied.

When measuring along pavements, levels should be taken at 50 m intervals on flat or even grades, at about 10 m intervals on vertical curves and always at points giving obvious vertical deviations such as kerbs. In fields, pegs or laths should be aligned at intervals of about 5 m and a nail driven in horizontally either to support or to bold down the band. These nails may be levelled in bays by eye and only those at the changes of gradient need be levelled by instrument.

Change points should be marked on the ground at intervals of a band length less about 200 mm. They may be set out by eye because the 100 m band is long compared with the probable error in their alignment. For example, should the alignment error be as high as 250 mm (obviously unlikely) then the error in measuring would not be greater than 1/320000.

The zero line on the band should be held firmly on the first ground-mark and not read against a scale (a very common source of mistakes). The band is tensioned to its standard, usually 5 kg (10 lb), and placed alongside the mark. The whole number of metres is then recorded and re-checked. The tension on the band is increased by about 1 kg, then re-set to the standard and read against the mark; the tension is then reduced by about 1 kg, re-set to the standard and read against the mark again. This procedure checks whether the band is running freely. Readings should be taken to the nearest millimetre only. Estimating more closely than this is not to be recommended for it has been found to create a false confidence in the accuracy of the measurement and so to preoccupy the observer with insignificant errors that he is prone to make mistakes of a whole metre or decimetre. Broadly speaking, readings to the nearest millimetre will lead to a maximum compensating error of plus or minus 0.5 mm giving a probable error in ten band lengths, for example, of about 1.6 mm or 1/620000. The same argument applies at the zero end of the band, where the setting error is probably plus or minus 0.3 mm.

Slope corrections are always negative and may be taken as the sum of $h^2/2L$, where h is the level difference in a given length L. It should be calculated to one millimetre only, greater accuracy is unnecessary and increases the likelihood of mistakes. Where slopes are steep, corrections may be calculated by trigonometry. The correction for horizontal deviation is calculated similarly.

The temperature correction for steel bands is $0.0000112Lt$, where L is the length measured in metres and t the difference in degrees Centigrade between the actual and standard band temperatures. The standard is normally 20°C (68°F). Where the actual temperature is less than standard, the correction is negative.

Band corrections, found by calibration against a standard length, may need to be applied and great care should be taken that they are applied in the correct sign.

Sea level corrections, always negative where the measured line is above Ordnance Datum, are applied to lengths measured at high altitudes.

The arbitrary sag/irregularities correction is always negative. It has been found by experience to vary from 0.5 mm (0.002 ft) per band length when measuring along pavements to 1.5 mm (0.006 ft) per band length when measuring across fields.

Although the corrections, other than for temperature, are generally small, they are all cumulative and therefore must always be taken into account in precise work.

Examples

1 FIELD BOOK

Station A - change mark 1   Cloudy, dry, pavement dry
Time ...   Date ...   100 × 7 mm band (6 mm too long)

Back	Inter	Fore	Chge.	h	L	$h^2/2L$	Remarks
2.14			0				Stn. A.
				0.74	7.0	0.039	
	2.88		7.0				
				1.52	24.0	0.048	
	1.36		31.0				
				2.51	26.3	0.119	
0.72		3.87	57.3				
				2.07	30.9	0.069	
	2.79		88.2				
				0.21	11.7	0.002	
		3.00	99.9				ch. mk. 1
			Correction		-	-0.277	slope

Ch.	deviation	d	L	$d^2/2L$
0	straight			
		0	-	0
28.0	straight			
		1.600	7.7	0.165
35.7	1.600 offset			
		1.600	14.3	0.089
50	straight			
		0	-	0
99.9	straight			
	Correction		-	-0.254 deviation

Temperature $6^{\circ}C$

$$-\frac{20}{14} \times 99.912 \times 0.0000112$$

$= -\ 0.016$ correction, temperature

Band   (6 mm too long)

$$+\ 0.006 \times \frac{99.912}{100}$$

$= +\ 0.006$ correction, band

Measurement			=	99.912
Corrections	+	-		
Slope		0.277		
Deviation		0.254		
Temperature		0.016		
Band	0.006			
	0.006	0.547	-	0.541
LENGTH STN. A - CH. MK. 1			=	99.371

Change marks 1-2 Theod. on (2). Ht. 1.410
Time ... Date ... 30 × 13 tape (held, one metre mark)

Vert. angle FL 102.00)
FR 257.20) $12^{\circ}.20'$

Temperature 6
20

Correction - 14 × 17.168 × 0.0000112 = -0.003 temp.

Measurement 17.168 = 17.168

18
-1
17

17.165

$17.165 \cos 12^{\circ}.20'$

1.2346438
9.9691734
1.2038172

= 15.989

LENGTH, CH. MK. 1-2 = 15.989

2 SUMMARY SHEET

STATION A TO STATION B

Corrections:

Irregularities/sag - 7 × 0.0005 = - 0.004

Sea level A: 583 O.D.) Average
B: 490 O.D.) 536 O.D.

$$- \frac{571}{1000} \times 536 \times \frac{0.078 + 0.094}{500 + 600} = - 0.048$$

TOTAL - 0.052

Measurement:	A-1	99.371
	1-2	15.989
	2-3	98.847
	3-4	99.590
	4-5	99.873
	5-6	99.143
	6-B	58.639
		571.452
Corrections		- 0.052
Station A - Station B		571.400 Horiz. Sea.Length

2 The Theodolite

The methods of observing and of arranging the field book which are set out in this chapter contain all that is normally necessary and sufficient for minimizing errors and obviating mistakes. They should be followed meticulously and never varied by the observer or between observers. On this basis, the whole of the survey work may be assumed to be of a consistent order of accuracy.

Horizontal Angles

Centre the theodolite over the station and level it approximately. With face left, orientate the instrument by sighting a station of known bearing with the top plate clamped and set to that bearing. Clamp the lower plate, unclamp the upper plate, and rotate the instrument about half a dozen times clockwise (swing right). From now on do not reverse this swing. Next, level the theodolite accurately and observe the stations in a clockwise direction through 360° sighting the same Referring Object (R.O.) both at the beginning and the end of these observations for the first pointing. The Referring Object is selected by choosing the clearest, remote station.

Should the observer overshoot any target, he should rotate the instrument through 360° without reversing its direction and approach the target again. The telescope is moved in the vertical plane by pressing it always downwards at one end or the other.

Micrometers should always be rotated in the same direction to obtain the reading and each reading checked three or four times for consistency. Changing face automatically reverses the "run" of the micrometer in alternate pointings.

With modern instruments it is not necessary to re-orientate it to read different parts of the circle. An older theodolite may require this however and the circle should be set to give the same values for the minutes and seconds when re-orientating the instrument to true bearing, followed by true bearing + 90°, + 135°, + 225°, for successive pointings. In this way, mistakes in individual readings may be checked quickly whilst observing, by comparing the minutes and seconds of the respective readings in successive pointings.

Change to face right for the second pointing. Keeping the lower plate clamped in the same orientation, rotate the instrument several times anticlockwise (swing left). Re-level and then observe, continuing always to swing left throughout this pointing.

These two pointings together constitute one round. The mean of them, a combination of the readings taken with opposite faces, gives the angles subtended by the observed stations. The seconds only of these mean values are entered in the field book in brackets immediately below the pair of pointings.

A second round is essential as a check upon possible mistakes in addition to increasing the accuracy of the observation. These two pointings should be observed swinging left with face left, then swinging right with face right, a different combination from the first round.

The mean values are entered in brackets as before and the difference between the first and second rounds noted at the foot of the page. Were it possible to measure all the angles with perfect accuracy, all these differences would be the same. In practice, they vary because of errors due to the instrument, changing atmospheric conditions, etc., and their variation is a measure of the accuracy of the set of observations. These are calculated in the field so that the accuracy of observing may be judged from them as well as from the closing errors in each pointing. Thus any need to re-observe may be seen before quitting the survey station.

Normally, four pointings will give sufficient accuracy, probably within two and a half seconds of arc when using a theodolite reading directly to one second.

The observed bearing is derived not from the mean values of the rounds (written in brackets) but is the average of all four pointings. It may be computed in the office.

No more than, say, fifteen stations should be observed in one set. Should there be more, they should be divided into two sets with four to six stations common to each. The common stations should be the clearest, most distant targets, distributed fairly evenly around the circle and one of them should be used at the Referring Object for both sets. Preferably, the two series of observations should be completed consecutively and on the same day.

Example of Field Book

At Extra — Slight breeze, overcast
Date ... — Time ... — Haze at Hercules, others clear
Observer — Instrument

Nineveh	Cuckoo	Hercules	Bury 4	R.O.	
257.44.43	302.25.52	199.30.06	229.09.18	44.45	FL 2
347.44.52	32.26.08	289.30.15	319.09.25	44.54	FR 2
(44.47)	(26.00)	(30.10)	(09.21)	(44.49)	(2)
32.44.56	77.26.14	334.30.17	4.09.26	44.58	FL 2
122.44.40	167.25.49	64.30.09	94.09.15	44.41	FR 1
(44.48)	(26.01)	(30.13)	(09.20)	(44.49)	(1)
257.44.48	302.26.01	199.30.12	229.09.21	44.49	1
+ 1	+ 1	+ 3	- 1	± 0	4

Note: The differences between bracketted figures are shown on the last line and indicate that the angles between Nineveh and Cuckoo agree exactly in each round, vary between Cuckoo and Hercules by two seconds, and so on. Their maximum variation, four seconds, is a measure of the accuracy of the complete observation.

Vertical Angles

Stations should be reciprocally observed wherever possible, the reciprocal readings being taken under similar atmospheric conditions at about the same time of day.

The best time to observe is between 1230 and 1530 when refraction is most nearly constant. Readings may also be taken at night after domestic heating has been reduced and before industry revives in the morning.

When plunging the telescope, it should be rotated by pressing down on the telescope, never by lifting it.

The vertical circle bubble must be re-set before observing each station. The targets must be observed with both faces and approached upwards and downwards alternately.

Example of Field Book

At Snowdon
Date ...   Time ...
Observer ...
Instrument ...

Occasional showers, overcast
Everest hazy, otherwise clear
Ht. of instrument 215
Ht. of target: bottom 480
Ht. of target: top 700

Ben Nevis		Everest		
87.41.12	~~359.59.48~~	107.17.27	~~360.00.14~~	FL
272.18.36	4.37.24	252.42.47	34.34.40	FR
(+2.18.42)		(-17.17.20)		
87.41.01	~~59.37~~	107.17.20	~~00.20~~	FL
272.18.36	4.37.35	252.43.00	34.34.20	FR
(+2.18.47)		(-17.17.10)		
87.41.09	~~59.41~~	107.17.24	~~00.12~~	FL
272.18.32	4.37.23	252.42.48	34.34.36	FR
(+2.18.41)		(-17.17.18)		
	6)13.52.22		6)103.43.36	
	+ 2.18.44		- 17.17.16	
			(Vane top)	
6"		10"		

In the field, the sum of each pair of pointings is written in the second column and checked for consistency to safeguard against mistakes and/or gross errors. These figures are then crossed out to avoid confusion. The rest of the arithmetic should be done in the office.

The difference between pointings is written in the second column and the resulting vertical angle, obtained by halving it, is carried back to the first column where it is identified by brackets. The total of the differences divided by six gives the observed vertical angle. Six pointings are taken to give an acceptable accuracy and because using a divisor of six is less likely to lead to arithmetical mistakes.

The final result of the observations should be checked against the bracketed values and their spread noted at the bottom of the page as an indication of the accuracy.

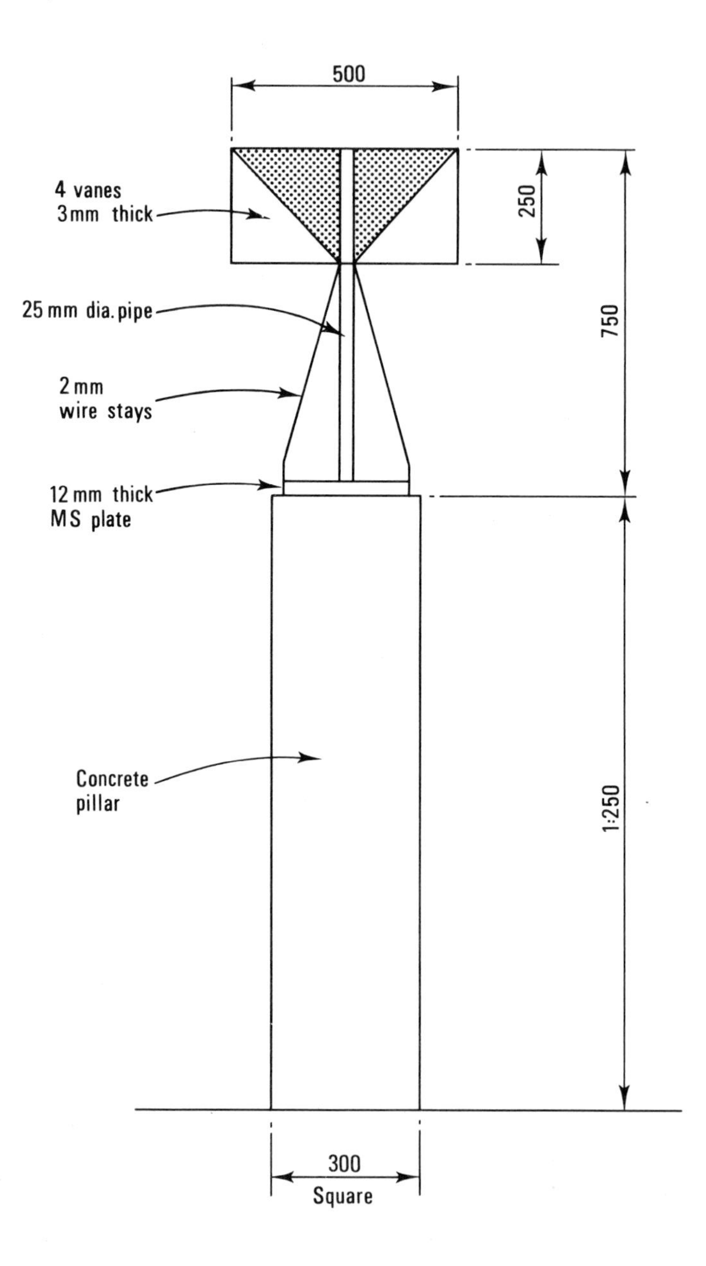

Figure 2.1 Survey Pillar, Instrument Stand and Target

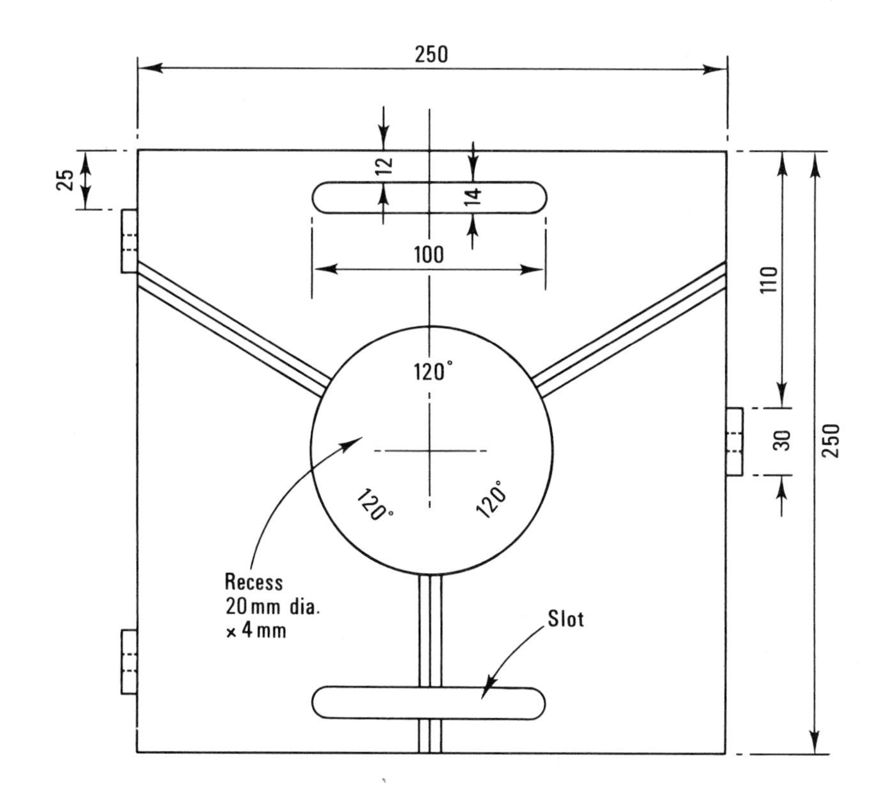

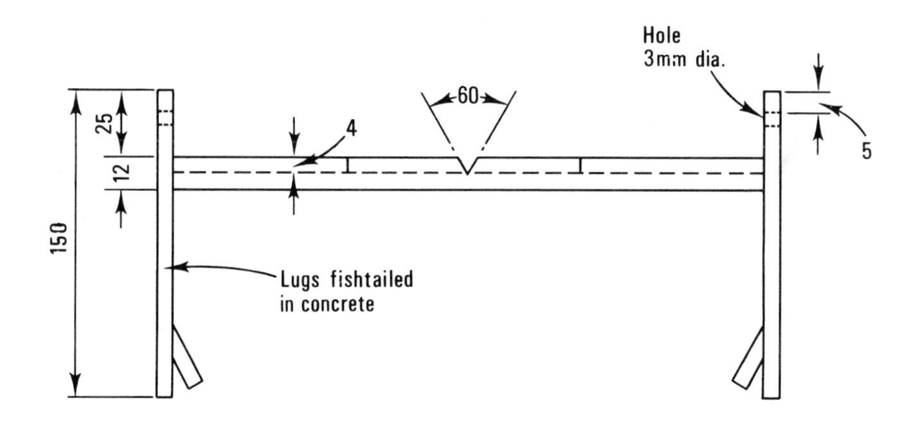

Figure 2.2 *Instrument and Target Stand set in Concrete Pillar*
Note: An adjustable stand may be made by using it without lugs, in conjunction with the base plate in Figure 2.3.

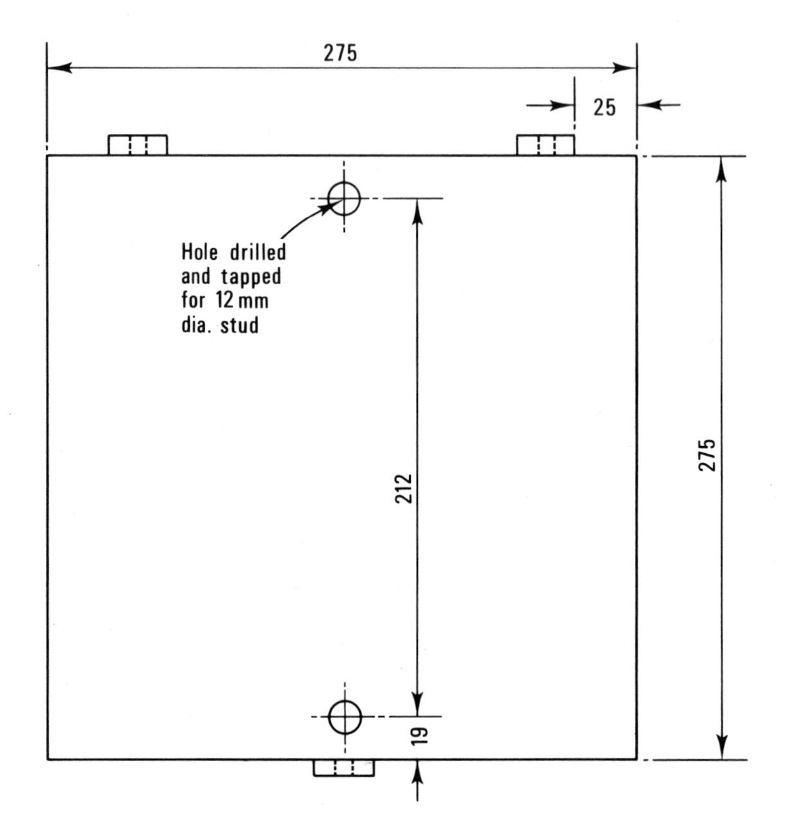

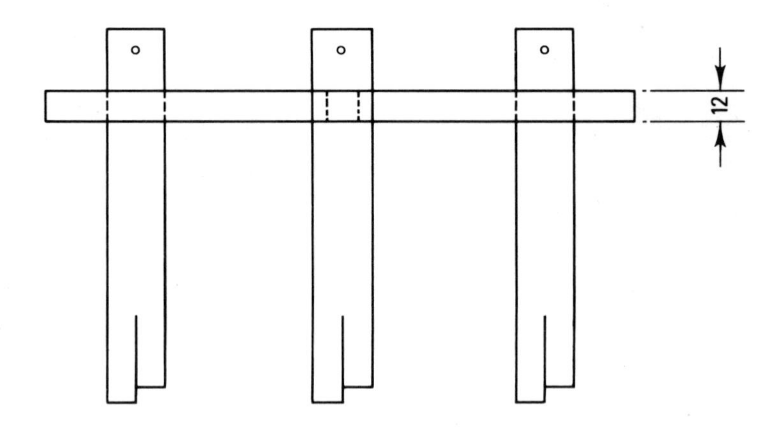

Figure 2.3 Baseplate for Adjustable Stand

3 Electromagnetic Distance Measurement

It has long been possible to measure with a very high degree of precision most physical quantities except length. Only comparatively recently, with the technological and technical advances in electronics, have we been enabled by EDM to measure distances in the field to an accuracy comparable with or better than that of angular measurement. This has revolutionized the approach to the production of accurate survey networks.

The several instruments which are available work on the common system of sending out radio or light rays which are returned to the initiating station by retransmission or reflection. The total time taken for the outward and return journeys is measured, and from it the distance between the stations is deduced. The original instruments in this field were the Geodimeter by Bergstrand and the Tellurometer by Wadley. Several variants of these have been produced by others, all based on the same general principles.

The Tellurometer

The Tellurometer, transmitting radio waves between two sets operated by two trained observers, offers many advantages for the survey of major networks and has obtained results of a high order of accuracy in national and international survey links. Its range is considerable, some 100 to 200 kilometres, and because the instruments allow speech communication also, the very considerable problem of liaison within a widely scattered survey organization is largely overcome. Readings may be obtained in haze, fog or rain, and darkness adds few difficulties to the fieldwork.

The possible errors in measurement are higher with the Tellurometer than the Geodimeter and may be up to ± 150 mm for the earlier models. This is being reduced continually by technical advance, but obviously gives a more acceptable proportionate error over the longer distance. The accuracy is affected where the radio waves pass close to the ground or between buildings, and the equipment cannot be used in the confined space of underground workings. The Tellurometer is at its best in the hands of the specialist surveyor engaged on principal frameworks. It is not described further in this book except that the forms of computation described in the section on geodimeter surveys are equally valid for the Tellurometer when used to measure comparable distances.

The Geodimeter

This instrument, requiring one operator, emits modulated pulses of light which are returned by a reflector system at the associated station. They are received by the Geodimeter which compares the difference in phase between the outgoing and the returned signals; hence the time delay due to the journey, and hence the distance between the instrument and the reflector. In simplified terms, polarized light is passed through a Kerr Cell, which is kept under controlled ambient conditions so that the long chain molecules of the cell rotate with a fixed frequency. The rotating molecules interrupt the passage of the light in a way similar to the action of a cine camera shutter. In parallel with the light emission, its phase is monitored by the circuitry where a controlled delay is superimposed. The delay is varied until a null-meter indicates that there is no phase difference between the returned light and the delayed pulse of the emitted light. At the same time, whether the phases are being compared in the positive half or the negative half of the cycle is disclosed by recording whether the delay dial knob and the null-meter pointer are both being turned in the same or opposite direction (S or O) whilst approaching the null point. Analagous to the elimination of theodolite errors by taking readings on both faces and at different positions on the circle, the delay line readings in the Geodimeter are taken at four different phase positions and their averages are compared. Variations within the instrument are monitored also by taking alternately, first the readings with the Geodimeter open, so that the light travels via the distant reflector, and then with it closed so that the light is reflected along a calibration path of fixed length within the instrument.

For a given frequency, a delay time covering one cycle corresponds to a certain distance, a delay of two cycles to twice that distance, and so on. Thus, for the same geodimeter reading there can be a series of lengths. The introduction of a different frequency will introduce "beats" which extends the period of this coincidence, and with more frequencies there is a further reduction of ambiguity.

The reflector system consists of one or more tetrahedral prisms which have the geometrical property of returning incoming rays of light along their original path, thus requiring the reflector setting to be only broadly in the direction of the Geodimeter (i.e. within about 15°). More prisms are used for long distances or strong daylight.

Normally, barometric and ambient temperature readings are taken at the Geodimeter during the observations. They may be recorded at the reflector station also and along the light path, but this is an unusual refinement. From these readings atmospheric corrections are applied to the measurement.

The author was responsible for the introduction and use in practice of the first Geodimeter purchased in the United Kingdom. Beyond the laboratory-based techniques for obtaining the slope distance between the instrument and the reflector, there had been no consideration as to methods for discovering its respective horizontal and vertical components, nor how to relate accurately these measurements to the survey station proper. There appeared to be no available expertise either in the choice of suitable frameworks, field techniques or methods of computation which would make the best use of the innovation. The systems which were evolved then by the author are described here and require minor changes only to take account of the technical improvements in the instruments.

Use of the Geodimeter

The various models have similar basic controls which are described here in general terms. Detailed information may be found for Model 4 Geodimeters in Appendix VIII and these notes may be amplified where necessary by obtaining the manufacturer's instruction sheet for the later designs.

Instrument checks are carried out first with the covers closed and the Lightpath Selector set for Calibrating, the Light switched off, and Instrument switches set for Adjusting or Testing. The Control Meter Selector is used to check in turn that the correct voltage and current are shown on the Control Meter, and then that the Null Indicator is correctly set at zero, if necessary, with the aid of the Zero Adjustment. The Light is then switched on and the Grey Wedge plunger for light sensitivity control is adjusted to obtain the correct current reading on the Control Meter.

Reflectors are set up at the associated station as described in the Appendix, or as required to suit the model of Geodimeter which is being employed.

In order to align the instrument, the front covers are opened, the Lightpath Selector opened to the Reflector, the correct Aperture chosen, and the Control Meter Selector set to energize the Null Indicator. The Coarse Sighting Telescope is used for approximate alignment, then the Eyepiece for closer adjustment, with the switch set for Adjusting, until the reflector at the associated station lies in the centre of the Kerr cell. For final alignment, with instrument switches set for Measuring and the Grey Wedge set to give the correct current reading on the Control Meter, the Geodimeter is adjusted for alignment so that a minimum current reading is obtained on the Control Meter. This adjustment and the readjustment of the Grey Wedge to give its correct current reading are repeated alternately until the best possible alignment is achieved.

For measuring, the Kerr cell is tuned afresh for each frequency with the Control Meter Selector set so that the Tuning Control obtains a minimum current reading on the Control Meter.

With the Frequency Selector set to the first frequency, the Lightpath Selector set for Calibrating, and the Control Meter Selector and switches set for Measuring, the Delay-line is adjusted to obtain a zero reading on the Null meter and the corresponding Delay Dial or Counter reading is recorded together with a note of whether they were rotating in the Same or Opposite sense; this is repeated for each of the Phase Selector positions.

The Lightpath Selector is opened to the Reflector and the above readings repeated.

The Frequency Selector is tuned to the next frequency, the readings repeated again, and so on until calibrating and reflector measurements have been recorded on all frequencies at each phase position.

Geodimeter Models

The maximum range at night of Geodimeter Models 4 and 6 is similar, about 15 km when using a tungsten lamp or 25 km with a mercury lamp, but the daylight range of the later model has been improved from 2 to 4 km using a tungsten and from 5 to 8 km with a mercury lamp. Its accuracy is increased also from ± (10 mm + 2 ppm) to ± (5 mm + 1 ppm).

The separate transmitting and receiving optical systems of the Model 4 require a selection of glass wedges to be introduced at the reflector in order to aim the returning light ray towards the opposite side of the instrument. This is not necessary for the Model 6 which has a combined coaxial optical system; it introduced also digital reading for the delay-line setting but requires still the conversion tables to give the distances. As with the Model 4, the distance must be known to within plus or minus one kilometre.

On the Model 6A, a resolver to give readings of the delay-line dial, which, in addition to having the counter for digital reading, are also linear with respect to distances, dispenses with the need for calibration tables. Furthermore, the use of four instead of three frequencies raises the limit of ambiguity in the measurements to 50 km.

The Geodimeter Model 8 introduces a laser light source which, being concentrated at the red end of the spectrum, is outside the range of most of the sun's radiation. The introduction of the narrow band filter, which is transparent to the laser but virtually opaque to the sun's rays, improves the signal to noise ratio. The range is about 60 km and much the same by day as by night. Because the laser beam is very concentrated, a "pointing" lens may be interposed during the initial alignment of the instrument in order to spread the beam vertically and so make easier the detection of the reflector station. Accuracy stands at about ± (5 mm + 1 ppm).

4 Levelling

Dumpey Levelling

The level traverse should be split into bays by establishing Temporary Bench Marks (T.B.M.) placed at intervals of not more than 500 m (quarter-mile) horizontally or 15 m (50 ft) vertically.

Each bay is levelled once in each direction with a maximum closing error of 5 mm (0.020 ft). Should any bay fail to close wtihin this limit, it may be broken into two parts by the introduction of an intermediate T.B.M. and each part levelled separately to close within the maximum permitted error of 5 mm.

An ordinary engineer's staff with 10 mm (0.01 ft) divisions is adequate but it should be checked from time to time by measuring with a steel tape to 1 mm (0.001 ft). A spirit level mounted on the staff is used to set it plumb; "swinging" the staff whilst reading it may introduce errors. Foresights and backsights must be made sensibly equal by training the chainman to walk an equal number of paces between the instrument and the change points. Steel change point plates incorporating three pointed feet for working on roads and hard ground or with a single spike about 100 mm long for work in fields are essential. The chainman should be instructed never to lift the staff from a change point but to rotate it carefully to face the instrument positions.

Any level in good adjustment should give acceptable results. Readings are estimated to 1 mm (0.001 ft) and the maximum sight in the United Kingdom should be about 50 m (150 ft). A tilting level is generally considered to be the most convenient but an engineer's level may be used in the same way as the quickset instrument without loss of accuracy. After levelling it approximately only, using the main and cross bubbles, the telescope is clamped and relevelled accurately by means of any one of the foots crews whilst observing the staff.

The outward and return runs for each bay should be booked on separate pages, the backsights and foresights added to give the total rise or fall, and the two runs compared to check the closing error. No further computation work should be attempted in the field.

In the office, these rises and falls are rechecked, tabulated on calculation sheets, and the sum of the outward runs compared with the sum of the return runs. The mean of these totals is then applied to the adopted Datum to obtain the reduced levels of all the T.B.M.

The traverse should be tied at intervals to the nearest convenient Ordnance Survey Bench Marks and this levelling carried out separately in independent bays. O.S. values shown on maps may be out of date, therefore a current Bench Mark list should be obtained from the Ordnance Survey. Even so, settlement may have introduced further discrepancies of several centimetres in areas liable to subsidence.

Example of Field Book

K5 ---------- K6

Date ...   Time ...

Instrument ...
Observer ...
Weather ...

Back	Inter	Fore	Coll	R.L.	Station
4.482					K5
4.293		4.178			
4.089		4.240			
4.527		3.984			
3.735		4.537			
		3.963			K6
+21.126		20.902			
-20.902					
+ 0.224	K5 -------- K6				

K6 ---------- K5

Date ...   Time ...

Instrument ...
Observer ...
Weather ...

Back	Inter	Fore	Coll	R.L.	Station
3.567					K6
4.238		3.352			
3.824		4.274			
4.192		3.930			
4.115		4.187			
		4.420			K5
+19.936		20.163			
-20.163					
- 0.227	K6 -------- K5				

Example of Calculation Sheet

Station	+	-	+	-
K8				
		0.292	0.296	
K7				
		0.350	0.345	
K6				
		0.227	0.224	
K5				
		0.267	0.266	
K4				
		0.218	0.223	
K3				
		2.015	2.019	
K2				
	1.884			1.879
K1				
	+ 1.884	- 3.369	3.373	1.879
		+ 1.884	1.879	
		- 1.485	1.494	
		- 1.494		
		2)- 2.279		
K8 - K1		- 1.489		
K8		144.853	(Datum)	
K1		143.364		

Note: The Datum should also be chosen (or modified by the application of, say, 100 or 1000 m or ft) so that all reduced levels are positive.

Trigonometrical Levelling

The heights of stations may be obtained with a very fair degree of accuracy by combining the horizontal distances between stations obtained from a triangulation survey, or the slope distances measured by geodimeter with vertical angles observed with a theodolite reading to one second of arc. Whilst Dumpey levelling gives good results when carried along the contours between points of about the same altitude, trigonometrical levelling gives a greater degree of accuracy where the stations are at considerably different altitudes and, more particularly, on steep slopes.

Observations
Vertical angles between stations should be observed reciprocally wherever possible, the reciprocal readings being taken under similar atmospheric conditions.
The method of observing them is described in the chapter on the theodolite.

Corrections

(1) Curvature and refraction correction

Where vertical angles have not been reciprocally observed, they must first be corrected to allow for the curvature of the earth and for terrestrial refraction.

The curvature correction is always positive.

Curvature correction = $+ L^2/2R$

where L is the horizontal distance between stations, and R the radius of curvature of the earth's surface about 6400000m (21000000 ft).

The refraction correction is always negative.

Refraction correction = $- kL^2/R$

The coefficient of terrestrial refraction, k, may be taken as 0.07 in Great Britain.

Combining these two corrections gives:

$$\text{Curvature and Refraction Correction} = + \frac{L^2}{R} \times \frac{1 - 2k}{2}$$

$$= + \frac{L}{R} \times \frac{1 - 2k}{2} \times 206265 \text{ seconds}$$

and results in the following table.

Horiz. Dist. (metres)	Correction (seconds)	Horiz. Dist. (metres)	Correction (seconds)
1000	14	6000	83
2000	28	7000	97
3000	42	8000	111
4000	56	9000	125
5000	69	10000	139

(2) Altitude correction

Where the angles are steep or the stations are at high altitudes, the horizontal distance between stations may need to be corrected to allow for their height above sea level before it is used to calculate their vertical difference.

The actual length L of a line joining two stations whose mean altitude is h feet is greater than their distance apart at sea level.

Altitude correction = $+ Lh/R$

The following table gives the altitude corrections for a horizontal sea level length of 1000 m.

Altitude (metres)	Correction (metres)	Altitude (metres)	Correction (metres)
100	0.016	600	0.094
200	0.031	700	0.109
300	0.047	800	0.125
400	0.062	900	0.140
500	0.078	1000	0.156

A distance measured with a band or by geodimeter is the actual length at that altitude. Consequently it needs no correction before being used to calculate the vertical distance. A description of this is given in the chapter on the geodimeter.

Computations

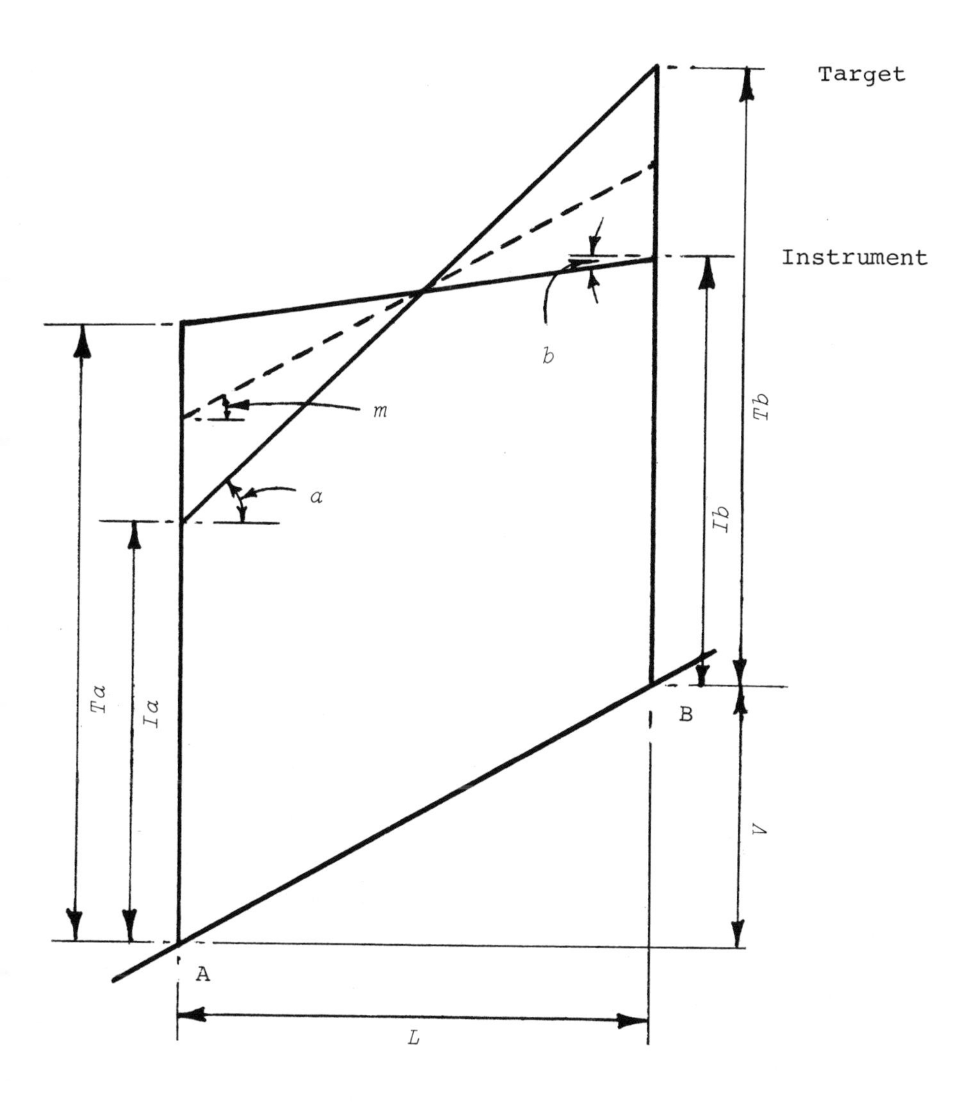

Figure 4.1

Let $\frac{1}{2}(a + b) = m$

Then $V = L\tan m + \frac{1}{2}(Ta + Ia) - \frac{1}{2}(Tb + Ib)$

This formula is not mathematically exact, but it is sufficiently accurate for practical purposes.

In this form it lends itself readily to the simple tabulation shown in the following examples.

Examples

OD				
6	JARROW	- 0.00.42	+ 1.520	+ 0.320
6	HOWDON	- 0.00.11	- 0.240	- 1.580
2)12		2)0.00.31	+ 1.280	- 1.260
6	2283.947	- 0.00.15	+ 0.640	- 0.630
0.001	0.002			+ 0.010
	2283.949	3.3586864		
		5.8616661		
		9.2203525		- 0.166
	Jarrow-Howdon			- 0.156

OD				
435	DAM	+ 28.30.15	+ 1.480	+ 1.900
770	CRUACHAN	- 28.28.01	- 1.420	- 1.570
2)1205		2)56.58.16	+ 0.060	+ 0.330
602	619.243	+ 28.29.08	+ 0.030	+ 0.165
0.094	0.058			+ 0.195
	619.301	2.7919019		
		9.7345033		
		2.5264052		+ 335.051
	Dam-Cruachan			+ 336.246

OD				
435	DAM	+	+	+ 1.900
770	CRUACHAN	- 28.28.01	- 1.420	-
2)1205		08		
602	619.243	+ 28.27.53	- 1.420	+ 1.900
0.094	0.058			+ 0.480
	619.301	2.7919019		
		9.7341265		
		2.5260284		+ 335.760
	Dam-Cruachan			+ 336.240

JARROW	Vert. angle	a	Ia	Ta
HOWDON	" "	b	Ib	Tb
	2) Difference		sum	sum
Sea length	Vert. angle	m	$\frac{1}{2}(Ia - Ib)$	$\frac{1}{2}(Ta - Tb)$
Alt. corr.			$\frac{1}{2}(Ta + Ia - Tb - Ib)$	
True length	True length			
	tan$m$			
	Vert. dist.		Vert. dist.	
	Jarrow-Howdon		V	

DAM				Ta
CRUACHAN	Vert. angle	b	Ib	
	Curv. refr. corr.			
Sea length	Vert. angle	m	$- Ib$	$+ Ta$
Alt. corr.			$- Ib + Ta$	
True length	True length			
	tan$m$			
	Vert. dist.		Vert. dist.	
	Dam-Cruachan		V	

5 Traverse Surveys

The traverse survey is normally employed in urban areas and, therefore, rarely is it possible to carry out the field work with sufficient accuracy by day. According to the neighbourhood, which may range from suburb to city centre, the measurement of bearings and lengths should be made between about 2200 and 0600.

This avoids the mutual conflict with people and vehicles, the vibration which may disturb the survey equipment, and the noise and movement which distracts the observer, whilst it permits almost complete freedom in the proper choice of instrument stations and lines of sight.

Stations should be as far apart as possible in order to reduce the angular error resulting from errors in placing the theodolite and targets over the survey points. It reduces also the total error which increases with the number of angles in the traverse. Therefore, the main framework frequently follows the main road rather than the back streets of the town, with secondary traverses based upon it linking and coordinating the points to be located.

It has been found that ambient temperature gradients across city streets cause horizontal refraction of the sight rays which may give rise to significant errors in angular measurements. These gradients are probably due to the heat, produced naturally and artificially, which emanates from the walls of buildings. Generally, poor results are obtained where lines run along the pavement; the best, where they are in the middle of the road. Angular errors may be caused also by sight rays passing over or close to illuminated street signs, sewer manholes, air conditioner outlets and other sources of heat.

Length is most easily measured along the centre of roadways. Their even gradient without irregularities and obstructions require fewer observations to establish the slope and deviation corrections, thus simplifying the calculations and also reducing the opportunities for mistakes.

As far as possible then, the framework should be of long lines joining stations placed near the centre of road junctions. It must be a closed system and should be strengthened where the opportunity arises by cross-links giving it a ladder formation. It may be strengthened further by tying the system to Ordnance Survey Stations or Revision Points.

There are occasions when an otherwise satisfactory framework of longlines is unavoidably interrupted by a short leg. It may be possible to overcome this by use of the Weisbach Triangle. The short leg is extended as far as possible in one or both directions and distant stations established which are sensibly in line with the short leg. By observing the distant stations in the traverse framework together with those on the extended line and by using the

Weisbach method, the bearing of the first long traverse leg is carried forward to the long extended line and from there to the succeeding long traverse leg. The bearing between the survey points defining the short leg may be observed also as a check but as its value is liable to greater error it should be ignored when adjusting the bearings of the main framework. Its bearing may be used, after adjustment as a secondary line, for the computation of the coordinates of the survey points on the short leg.

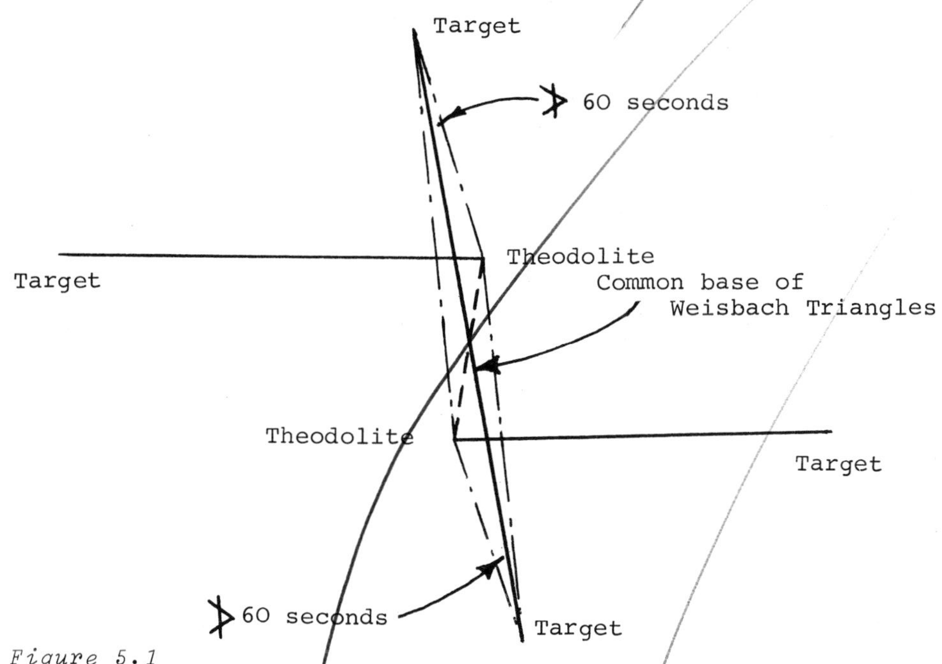

Figure 5.1

The calculations for the Weisbach method are similar to those described in the chapter on shaft plumbing.

Should it be impossible to extend the short leg in this way, a plane mirror (or mirrors) may be used to align imaginary distant stations forming the common base line for the Weisbach Triangles. The observations at both survey stations are made consecutively without the mirrors' being moved in the meantime.

The agreement obtained in a carefully executed closed traverse often makes the adjustment of errors in bearings and lengths a matter almost of academic interest. The adjustment is carried out mainly to produce an arithmetically correct primary network on to which the secondary traverses may be closed more readily.

It is argued sometimes that the high order of agreement which is obtained so frequently makes the use of the closed traverse unnecessary. This argument is never valid. Obviously, an open traverse carries no guarantee of accuracy, gives no warning of errors due to undisclosed field effects on the measurements, and it guards in no way against mistakes during observation or against mistakes in the computation.

Stations may be established by driving a woodscrew with a cruciform cut head into the road surfacing or between (but not within) paving slabs. Alternatively, chiselled crosses or punched centre marks may be made on the frames of manholes covers and road studs.

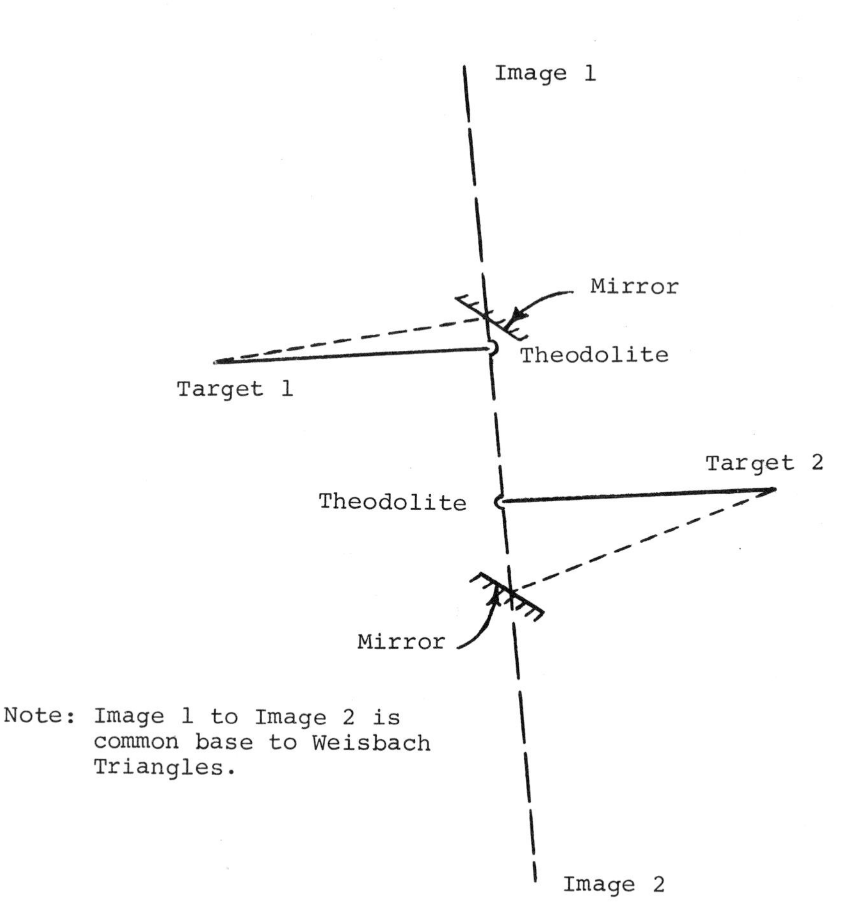

Figure 5.2

Light-boxes, the front face screened by tracing paper and illuminated from behind by an incandescent pressure lamp, effectively silhouette a fine line and plumb bob for short sights. For long distances, say 300 to 2000 m, an illuminated vertical slit may be formed by two pieces of thick cardboard pinned on top of the tracing paper; the width of the slit is varied to suit the length of sight and normally it should subtend an angle of about four seconds of arc at the instrument. For greater distances, the incandescent lamp may be set accurately above the survey station and its mantle observed directly. The high-temperature light source enables the targets to be observed clearly even when quite foggy conditions prevent their being seen with the naked eye.

Normally, angular measurements are made with a theodolite reading to one second of arc. On pavements, the tripod should be set on the paving stones and not in the joints between them. These slabs must be checked to ensure that they do not rock, and chosen so that the observer may stand at and move round the instrument without stepping upon them.

An adequate number of red lights and warning signs are essential to protect the survey team and equipment. They must be placed with care so that the columns of warm air rising from the lamps do not affect the sight rays.

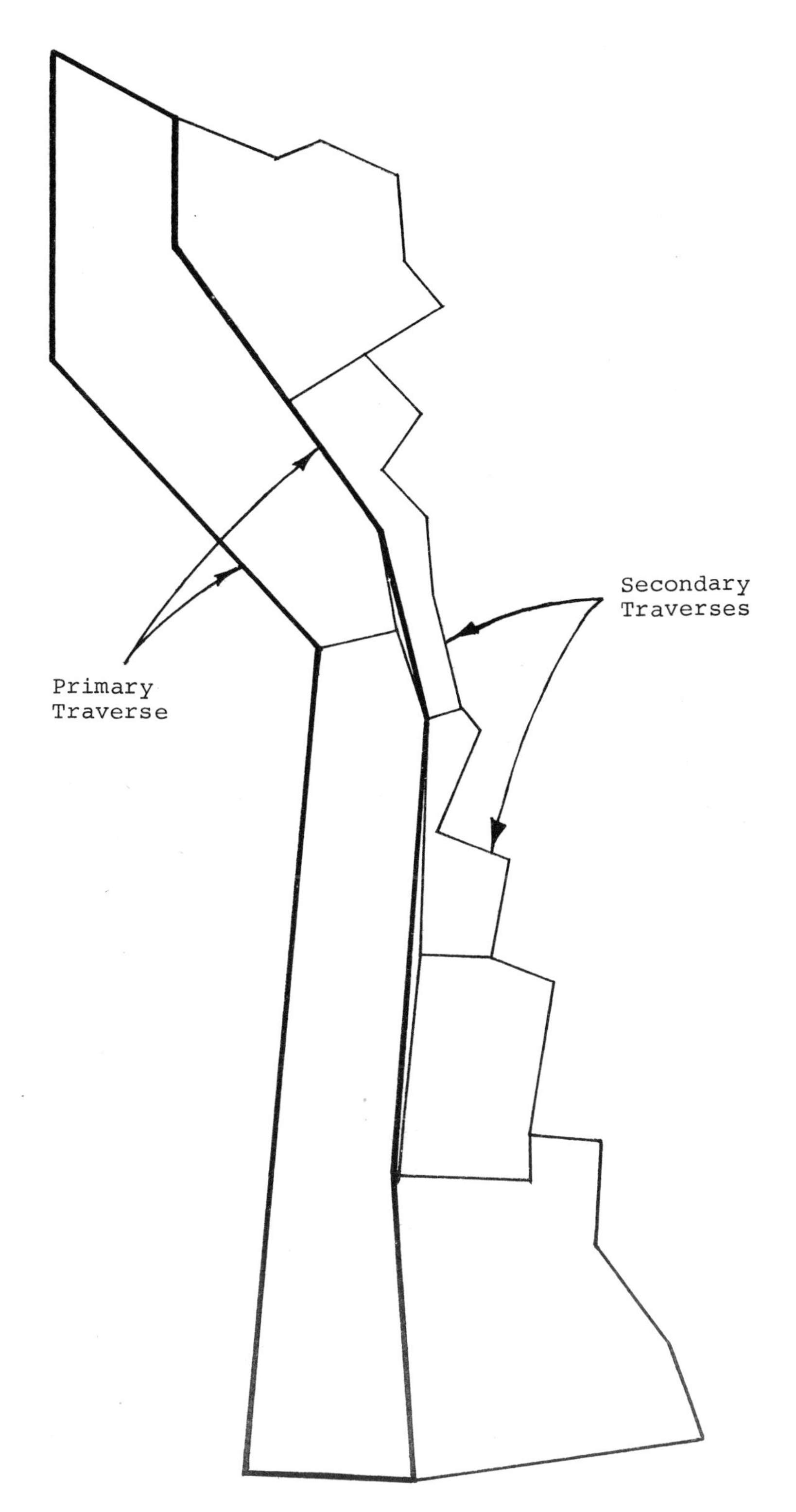

*Figure 5.3 Traverse Framework
Scale 1/10000*

The techniques and methods described in the chapters on the use of the band and the theodolite provide all that is necessary and sufficient for the linear and angular measurements in a traverse framework.

It should be possible to complete three or four angular observations or the measurement of lines totalling about 1000 to 1500 metres in length in one night. Beyond that, accuracy suffers because of mental and physical fatigue.

Using a single second theodolite, the angular closing error in a circuit consisting of N angles should not exceed $2.5\sqrt{N}$ seconds. This error is assumed to be distributed evenly throughout the framework and normally the corrections are applied more to the bearings of the short lines than to those of the longer ones. The corrected bearings are applied to the measured lengths to derive the respective latitudes and departures.

The closing errors are frequently of the order of 1 in 50000 or better with respect to linear measurement within the traverse. This is an indication of the agreement between the several measurements but not of their accuracy. Many errors will have been compensating; others, although cumulative, will have accumulated at much the same rate and, therefore, over the outward and return routes in a closed traverse, may have largely cancelled themselves. Nevertheless, the accuracy of the traverse overall is affected still by the accumulated errors. Where a framework is closed on to Ordnance Survey stations, for example, it may well be confirmed that the accuracy, as opposed to the internal agreement, of the traverse is about 1 in 20000.

It is desirable to retain the corrected bearings of the lines when the next step, the adjustment for the closing errors of the latitudes and departures, is carried out. Crandall's method is a suitable one and lends itself readily to a standard form of calculations. The proof is set out in Appendix VII.

Example of Traverse Computation

Table 1. Measured Lengths and Adjusted Bearings

Station	Measured Length (S)	Measured Bearing	Bearing Correction	Adjusted Bearing (A)
Oxford Peter				
	805.394	71.12.29	0	71.12.29
York Peter				
	381.804	250.38.18	- 1	250.38.17
Brook Peter				
	555.340	164.31.58	- 2	164.31.56
Brook Whitworth				
	367.599	106.10.20	- 3	106.10.17
London Whitworth				
	859.049	196.20.14	- 4	196.20.10
Brook Andrew				
	1128.971	275.06.31	- 4	275.06.27
Sale Andrew				
	651.691	95.39.21	- 5	95.39.16
Oxford Andrew				
	1306.044	351.49.30	- 5	351.49.25
Oxford Peter				
	-	71.12.35	- 6	71.12.29
York Peter				
Bearing error		+0.00.06		

From Crandall's Method, the corrections l and d for the latitude Y and departure X respectively of a line of length S and bearing A are then:

$l = fY\cos A + gS\cos A\sin A$

$d = fS\cos A\sin A + gX\sin A$

which may be developed thus:

STATION P - STATION Q			Logarithms	
$(2) + (5) =$	(7)	$Y\cos A = S\cos^2 A$	
$(1) + (2) =$	(5)	$S\cos A = Y$	
	(2)	$\cos A$	
Bearing A		(1)	S	
Length S		(3)	$\sin A$	$S\cos A\sin A$ $(1) + (2) + (3) = (4)$
$(1) + (3) =$	(6)	$S\sin A = X$	
$(3) + (6) =$	(8)	$X\sin A = S\sin^2 A$	
$(7) + (8) =$	(9)	$S^2\cos^2 A\sin^2 A$	
$\frac{1}{2} \times (9) =$	(10)	$S\cos A\sin A = (4)$ (check)	

$Y(5)$ $X(6)$

$Y\cos A$ (7) $S\cos A\sin A$ (4) $X\sin A$ (8)

The numbers in brackets indicate the order in which the computation is made.

OXFORD PETER - YORK PETER

	1.9220779	
	2.4140432	
71.12.29	9.5080347	
805.394	2.9060085	
	9.9762099	2.3902531
	2.8822184	
	2.8584283	
	5.7805062	
	2.8902531	

+ 259.444 + 762.462

+ 83.575 + 245.614 + 721.819

YORK PETER - BROOK PETER

	1.6228981	
	2.1023693	
250.38.17	9.5205288	
381.804	2.5818405	
	9.9747157	2.0770850
	2.5565562	
	2.5312719	
	5.1541700	
	2.5770850	

- 126.581 - 360.210

+ 41.966 + 119.422 + 339.838

BROOK PETER - BROOK WHITWORTH

```
                          2.7125155
                          2.7285373
164.31.56                 9.9839782
555.340                   2.7445591
                          9.4260172        2.1545545
                          2.1705763
                          1.5965935
                          5.3091090
                          2.6545545

      - 535.226 + 148.107

+ 515.841   - 142.743   + 39.500
```

Similarly, the coordinate differences and correction factors are derived for the remainder of the lines in the traverse.

Table 2. Differences of Coordinates (uncorrected)

Station	Y +	Y -	X +	X -
Oxford Peter				
	259.444		762.462	
York Peter				
		126.581		360.210
Brook Peter				
		535.226	148.107	
Brook Whitworth				
		102.381	353.054	
London Whitworth				
		824.368		241.626
Brook Andrew				
	100.506			1124.489
Sale Andrew				
		64.210	648.520	
Oxford Andrew				
	1292.768			185.747
Oxford Peter				
	1652.718	1652.766 1652.718	1912.147 1912.072	1912.072
Total Error		- 0.048	+ 0.071	
Total Correction		+ 0.048	- 0.071	

Table 3. Correction Factors

Station	Y^2/S +	YX/S +	YX/S -	X^2/S +
Oxford Peter				
	83.575	245.614		721.819
York Peter				
	41.966	119.422		339.838
Brook Peter				
	515.841		142.743	39.500
Brook Whitworth				
	28.514		98.330	339.085
London Whitworth				
	791.086	231.871		67.963
Brook Andrew				
	8.948		100.107	1120.024
Sale Andrew				
	6.327		63.898	645.365
Oxford Andrew				
	1279.627		183.859	26.417
Oxford Peter				
	2755.884	596.907 588.937	588.937	3300.011
		+ 7.970		

By Crandall's Method:

$$+ 0.048 = + 2755.884f + 7.970g$$

$$- 0.071 = + 7.970f + 3300.011g$$

Hence

$$f = + 0.0000175$$

$$g = - 0.0000216$$

Then, by applying f and g to Table 3 by slide rule we obtain Table 4.

Table 4. Corrections

Station	l		d	
Oxford Peter	+ 0.001 - 0.005	- 0.004	+ 0.004 - 0.016	- 0.012
York Peter	+ 0.001 - 0.003	- 0.002	+ 0.002 - 0.007	- 0.005
Brook Peter	+ 0.009 + 0.003	+ 0.012	- 0.002 - 0.001	- 0.003
Brook Whitworth	+ 0.001 + 0.002	+ 0.003	- 0.002 - 0.007	- 0.009
London Whitworth	+ 0.014 - 0.005	+ 0.009	+ 0.004 - 0.001	+ 0.003
Brook Andrew	+ 0.000 + 0.002	+ 0.002	- 0.002 - 0.024	- 0.026
Sale Andrew	+ 0.000 + 0.001	+ 0.001	- 0.001 - 0.014	- 0.015
Oxford Andrew	+ 0.023 + 0.004	+ 0.027	- 0.003 - 0.001	- 0.004
Oxford Peter				
Total Correction Total Error (check Table 2)		+ 0.048 - 0.048		- 0.071 + 0.071

In the above tables, for Brook Peter to Brook Whitworth, for example:

Correction to difference in Y coordinates (l)

$$= + 0.0000175 \times 515.841 - 0.0000216 \times (-142.743)$$
$$= + 0.009 \qquad + 0.003$$
$$= + 0.012$$

Correction to difference in X coordinates (d)

$$= + 0.0000175 \times (-142.743) - 0.0000216 \times 39.50$$
$$= - 0.002 \qquad - 0.001$$
$$= - 0.003$$

The final, adjusted coordinates of the stations are derived by combining Tables 2 and 4 and applying them to the grid coordinates:

Table 5. Final Coordinates

Given Oxford Peter + 5000.000 + 2000.000

Station	Y	X
Oxford Peter	+ 5000.000 + 259.444 - 0.004	+ 2000.000 + 762.462 - 0.012
York Peter	+ 5259.440 - 126.581 - 0.002	+ 2762.450 - 360.210 - 0.005
Brook Peter	+ 5132.857 - 535.226 + 0.012	+ 2402.235 + 148.107 - 0.003
Brook Whitworth	+ 4597.643 - 102.381 + 0.003	+ 2550.339 + 353.054 - 0.009
London Whitworth	+ 4495.265 - 824.368 + 0.009	+ 2903.384 - 241.626 + 0.003
Brook Andrew	+ 3670.906 + 100.506 + 0.002	+ 2661.761 - 1124.489 - 0.026
Sale Andrew	+ 3771.414 - 64.210 + 0.001	+ 1537.246 + 648.520 - 0.015
Oxford Andrew	+ 3707.205 + 1292.768 + 0.027	+ 2185.751 - 185.747 - 0.004
Oxford Peter (check)	+ 5000.000	+ 2000.000

All final bearings and lengths are deduced from the final coordinates in Table 5 and these values are the ones used for extension into secondary traverses or for any survey control work. The final bearings and distances are compared with the adjusted bearings (which should be the same) and with the measured lengths which should contain reasonably consistent errors.

For example:

BROOK WHITWORTH - LONDON WHITWORTH

+ 4597.643	+ 2550.339	2.0102069
+ 4495.265	+ 2903.384	2.5478301
- 102.378	+ 353.045	9.4623768
		9.9824670
		2.5653631

106.10.17 (106.10.17 Adj.)

367.589 (367.599 Meas.)

OXFORD ANDREW - OXFORD PETER

+ 3707.205	+ 2185.751	3.1115297
+ 5000.000	+ 2000.000	2.2689311
+ 1292.795	- 185.751	0.8425986
		9.9955628
		3.1159669

351.49.25 (359.49.25 Adj.)

1306.072 (1306.044 Meas.)

Giving comparable corrections of

- 0.010 in 367 or 1/36700

and + 0.028 in 1306 or 1/46700

6 Triangulation Surveys

Normally, triangulation surveys are used in rural areas.

The proper reconnaissance for the design of the framework joining the survey stations is the key to their accurate coordination. The cardinal rule, as with all surveying, is to consider first the major network, paying a passing regard only to the particular locations which are to be linked by the survey. It is essential that the primary network should consist of well conditioned triangles, all of similar proportions, their lines of sight passing well clear of the ground and obstructions. As far as may be determined, sight rays should not pass over towns or other areas which would be likely to produce rising columns of warm air which might introduce lateral refraction. The main framework should extend beyond and embrace the locations to be served by it.

Having adopted a suitable primary network, it may be extended then through equally well designed and reconnoitred secondary and tertiary frameworks to survey the particular features to be located.

Triangulation is probably most easily applied in truly country areas which are wild and bare. Difficulties arise in the civilised countryside where man-made forests, hedges, buildings may modify the well conditioned geometry of the system. Here, the geodimeter is of considerable value because it allows the introduction of some of the attributes of the traverse survey or trilateration to strengthen a weakened triangulation framework.

This chapter concerns itself with a system based purely on triangulation but the reader should supplement it by referring to those on the traverse and the geodimeter.

The Base

Ordnance Survey Triangulation Stations provide the simplest and strongest basis for such a framework but base lines may be measured with a band or a geodimeter.

Ordnance Survey coordinates for the National Grid are on a Transverse Mercator Projection, they are in metres and are large. For convenience they should be reduced by adopting a local origin a little to the south and west of the area to be surveyed; they must be converted from the plane of the National Grid Projection to sea level and if desired may be converted from metres to feet. National Grid coordinates specify Eastings first, Northings second but in the computations described below which have been designed to suit Shortredes Tables, Y (North) is always given first, X (East) second.

Example

Approx. centre of local survey   507 km Easting

From the O.S. Pamphlet on the Traverse Mercator Projection:

	LocalLocal	Scale	Factor	for	510 km E	=	0.99975
change in	"	"	"	"	500 km	=	0.99972
"	"	"	"	"	10 km	=	0.00003
"	"	"	"	"	7 km	=	0.00002
"	"	"	"	"	507 km	=	0.99974

Factor for converting National Grid coordinates in metres to Local Grid coordinates at sea level is

$$\frac{1}{0.99974}$$

	9.9998871
	0.0000000
1.00026	0.0001129

Adopt Local Origin 168 km North, 503 km East in order to make positive the coordinates of all points in the local survey.

LONDON AIRPORT CONTROL TOWER (rivet)

Y (Northings)	*X* (Eastings)	
+ 175762.39	+ 507531.78	metres N.G.
- 168	- 503	
+ 7762.39	4531.78	
3.8899954	3.6562689	
0.0001129	0.0001129	
3.8901083	3.6563818	
+ 7764.407	+ 4532.959	metres Local Grid

Observations

For major surveys a theodolite reading to one second of arc is essential.

To facilitate the location of stations in the field and to suit the method of computing, the bearing is carried forward from station to station, i.e. the theodolite circle is set so that the bearings recorded during observing lie within 20 to 30 seconds of the grid bearings and require only a small arithmetical correction before being used.

A description of the use of the instrument is set out in the chapter on the theodolite.

General Description of the Computations

The method described here, an unusual one, was evolved by C.T. Cogle whilst practising as a licensed surveyor in the then Colonies. It was used on some Colonial surveys and for the control of deep mine projects on the Rand, and it has since been used successfully on several major civil engineering projects in the United Kingdom.

Basically, the framework consists of braced quadrilaterals except that extra, cross-linking, stiffening rays are also observed

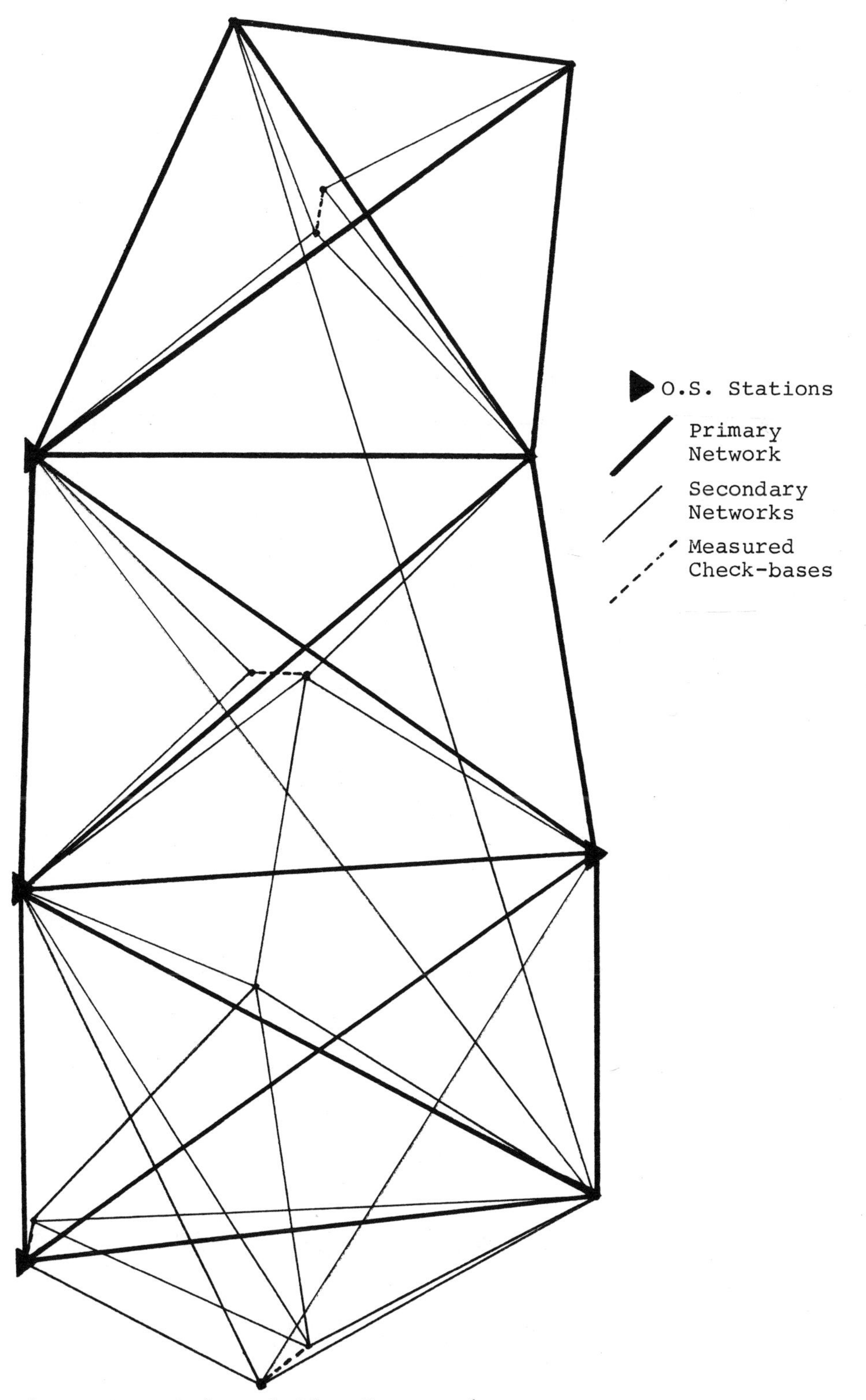

Figure 6.1 Triangulation Framework Scale 1/50000

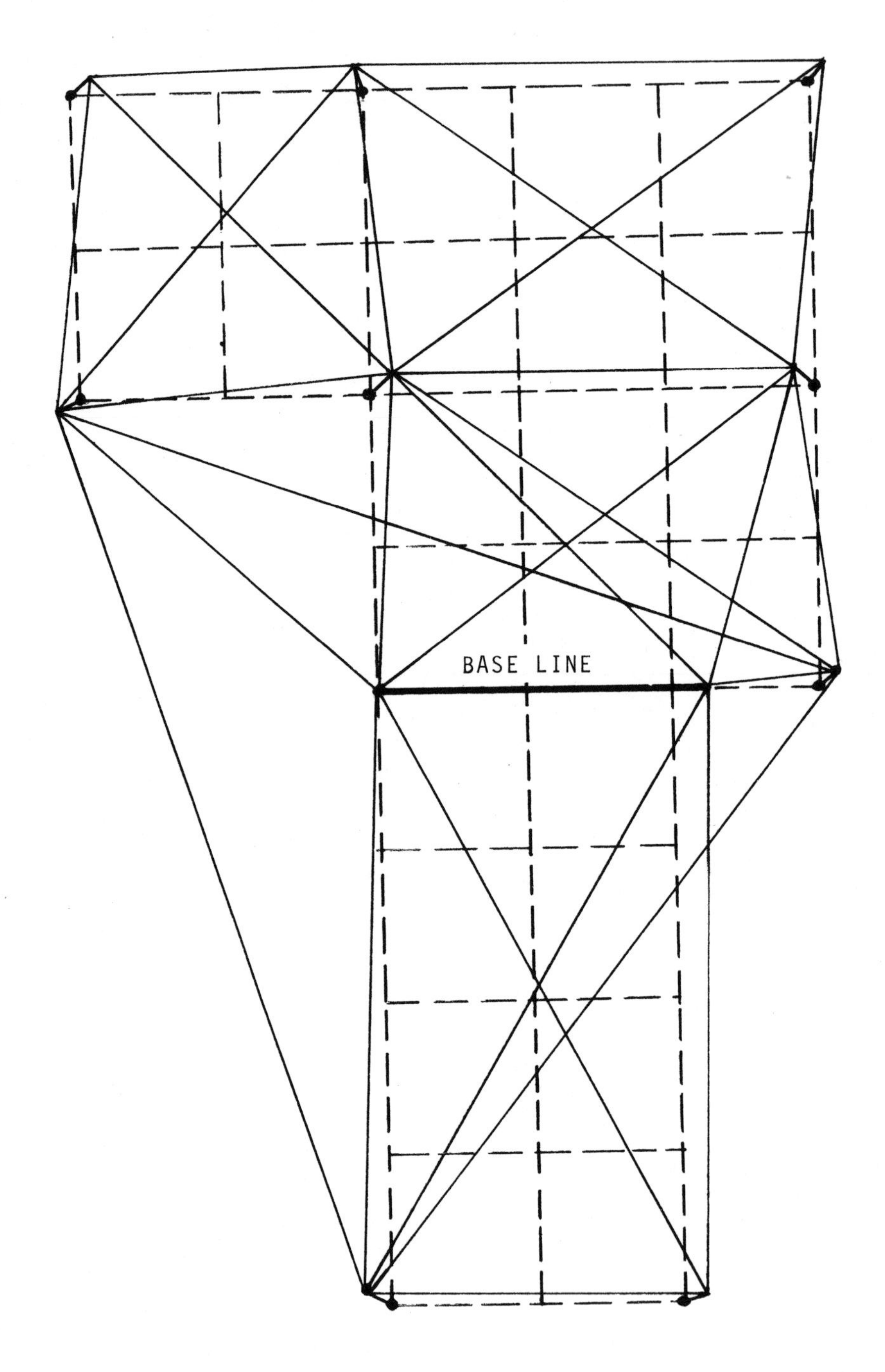

Figure 6.2 Triangulation Control for Rectangular Grid Scale 1/5000

wherever possible, analogous to the redundant members in certain structural frameworks.

All of these rays are used to determine the position of a station. They are plotted on graph paper, generally full scale, and the final coordinates of the station are deduced by inspection. Thus "sport" observations may be discounted or re-observed if necessary; the accuracy of the observations is immediately apparent and the error from a faulty reading is not carried forward unknowingly to the detriment of the sound observations.

Properly used the system is fully self-checking and may also be commended in that it calls only for a knowledge of the elements of trigonometry and a measure of commonsense.

Computations

The following example is of a survey based upon Ordnance Survey Stations. The lengths and bearings of the lines joining the O.S. Stations are calculated and these form the base lines for the triangulation.

Local Grid coordinates, at sea level, derived from National Grid Coordinates are:

	Y (North)	*X* (East)
Cribden Hill	+ 14040.100	+ 4948.760
Great Hameldon	+ 18932.200	+ 4416.740
Peel Park	+ 19370.430	+ 2235.370

CRIBDEN HILL - GREAT HAMELDON				
+ 14040.100	+ 4948.760	3.6894953	4920.944	4920.944
+ 18932.200	+ 4416.740	2.7259280	08314 990	94118 010
+ 4892.100	- 532.020	0.9635673	4428 850	492 094
353.47.36	4920.944	9.9974468	442 885	39 367
		3.6920485	19 684	492
Bearing	Length		492	49
			148	20
			39	4
			4892.098	532.026

Similarly:

Cribden Hill - Peel Park	333.01.18	5981.223
Great Hameldon - Peel Park	281.21.34	2224.955

In the above calculation, the coordinates are written down in the same order as the title. Their differences are written underneath and the respective signs of the differences are given by the mnemonic "top sign changes'. These signs are important as they are used later to identify the whole circle bearing.

The logarithms of these differences are taken down in their written order. Their difference is the cotangent or tangent of the bearing (which one is of no interest) and it will be found in the second or third column in Shortrede's Tables, i.e. to the left or right of the centre of the page. The tables then select the correct whole circle bearing.

In the example:

Signs of differences:	+ first	- second
Bearing found on the:	right in the Tables	
Locating always:	cosine first,	sine second.

The correct bearing is the one where the cosine on the right is positive and where the sine on the left is negative, namely 353°47'36".

The sign, + or - and the position, right or left, will appear in different combinations depending on the coordinates; the mnemonic "cos first, sin second", helps to select the correct bearing from the sixteen possibilities.

The next logarithm (9.9974468 in the example) is taken always from the fourth column of Shortrede's Tables. It is the cosine or sine of the bearing (which one is of no interest). This figure is subtracted always from the larger of the two lengths above to obtain the distance from station to station.

The check by natural numbers is carried out by applying the length to the cosine and sine of the bearing (cos first, sin second), This gives the coordinate differences in the correct order, *Y* first, *X* second.

To obtain answers correct to a given significant figure, the cosine and sine are written backwards under the length, their unit figure being written immediately under the significant figure required. Inspection of the example shows how the answers have been obtained correct to three decimal places.

Where the framework is built upon a measured base line, coordinates are assumed for a station at one end of it and a bearing assumed for the base line. The following calculation gives the coordinates of the station at the other end.

Example

BASE E - BASE W			
		3.0238160(*Y*)	(Naturals check as above)
240°00'00"		9.6989700 (cos first)	
2112.740		3.3248460 (length)	
		9.9375306 (sin second)	
		3.2623766(*X*)	
	- 1056.370	- 1829,690	(Tables give signs)
Base E	+ 5000.000	+ 2000.000	
Base W	+ 3943.630	+ 170.310	

Having established the true bearings between the known stations by the above methods, the readings from the field books may be orientated and adjusted.

AT CRIBDEN HILL - 6"

	Field Book	Adjusted Bearings
Haslingden Moor	261°58'48"	261°58'42"
Pen Moss	325.19.14	325.19.08
Peel Park	333.01.24(-6")	
Hameldon Eccentric	353.21.10	353.21.04
Great Hameldon	353.47.43(-7")	
Goodshaw Hill	354.16.03	354.15.57

AT GREAT HAMELDON [- 13"]

Goodshaw Hill	173.08.42	173.08.29
Cribden Hill	173.47.49(-13")	
Haslingden Moor	201.25.53	201.25.40
Pen Moss	201.42.53	201.42.40
Hameldon Eccentric	240.35.55	240.35.42
Peel Park	281.21.47(-13")	

AT PEEL PARK [+ 9"]

Haslingden Moor	178°48'25"	178°48'44"
Great Hameldon	101.21.26(+8")	
Hameldon Eccentric	102.03.36	102.03.45
Goodshaw Hill	135.42.29	135.42.38
Cribden Hill	153.01.08(+10")	
Pen Moss	159.32.48	159.32.57

AT GOODSHAW HILL [+ 2"]

Cribden Hill	174.15.57(±0")	174.15.58
Haslingden Moor	215.45.55	215.45.57
Pen Moss	246.16.34	246.16.36
Peel Park	315.42.36(+2")	315.42.38
Hameldon Eccentric	352.04.36	352.04.38
Great Hameldon	353.08.26(+3")	353.08.29

AT HASLINGDEN MOOR [+ 19"]

Goodshaw Hill	35°49.39(+18")	35.45.57
Cribden Hill	81.58.22(+20")	81.58.42
Peel Park	358.48.26(+18")	358.48.44
Hameldon Eccentric	21.09.20	21.09.39
Great Hameldon	21.25.18(+22")	21.25.39

AT HAMELDON ECCENTRIC [+ 10'.17"]

Goodshaw Hill	171.54.19(+10'19")	172.04.37
Cribden Hill	173.10.46(+10'18")	173.21.04
Haslingden Moor	200.59.26(+10'13")	201.09.41
Pen Moss	201.00.23	201.10.40
Peel Park	281.53.25(+10'20")	282.03.44
Great Hameldon	60.25.24(+10'18")*	60.35.42

*Short line, ignore in average

AT PEN MOSS [- 9"]

Great Hameldon	21.42.51(-11")	21.42.41
Goodshaw Hill	66.16.43(-7")	66.16.35
Cribden Hill	145.19.14(-6")	145.19.07
Haslingden Moor	201.08.54 (one direction)	201.08.45
Peel Park	339.33.07(-10")	339.32.57
Hameldon Eccentric	21.10.52(-12")	21.10.41

Note: At Cribden Hill, the bearing to Peel Park requires a correction of (- 6"), that to Great Hameldon (- 7"); the average (to the nearest second). [- 6"] is applied to the Fieldbook readings to give the adjusted values. These are now carried forward to their respective reciprocal observations. Goodshaw Hill refers to the adjusted bearings to it from Cribden Hill, Peel Park and Great Hameldon.

At Goodshaw Hill, then, the average adjustment is [+ 2"] and in correcting the bearing to, say, Cribden Hill, the mean of (± 0") and [+ 2"] is applied, a final adjustment of + 1" which takes account of the incoming and outgoing rays. No other observations affect it and this final value is underlined.

The method of observing and booking checks the accuracy of each individual reading. The successive averaging of the results throughout the first part of the computations not only makes the best of the observations but also provides a further check on accuracy by setting out the (bracketed) corrections for inspection of their variation. Should there be poor agreement, the incoming and outgoing rays may be plotted separately.

Adjusted bearings not underlined are tentative; those underlined are final and are the ones used in the calculations for the plotting.

In the given example, besides the O.S. Stations, the following stations had been computed previously:

Pen Moss	+ 16315.030	+ 3374.670
Hameldon Eccentric	+ 18911.960	+ 4380.830
Haslingden Moor	+ 13674.350	+ 2353.420

The next step is to compute a "provisional" value for the new point by applying the sine rule to a triangle formed by the new point and any two of the known points. The bearing and length of the line joining the two known beacons is derived from their coordinates; the bearings from them to the unknown beacon are given by the observations and from these bearings the internal angles of the triangle may be deduced. This calculation is self-checking, an important advantage.

In order to suit the arrangement of Shortrede's Tables, the sine rule calculation is set out in clockwise order around the triangle.

Referring to Figures 6.3 and 6.4, the bearing and length of the known base is calculated in the clockwise direction Pen Moss to Peel Park.

Then the three stations are written in clockwise order with the unknown station in the middle of them:

Peel Park
Goodshaw Hill
Pen Moss

For convenient manipulation, the length and bearing of the known base is interposed:

Peel Park
Pen Moss - Peel Park
Goodshaw Hill
Pen Moss

The title "Pen Moss - Peel Park" also indicates that, where the calculation is repeated or duplicated, the left-hand calculation relates to Pen Moss, the right hand calculation to Peel Park.

On the line "Pen Moss - Peel Park" is written the logarithm of its length followed by the bearings Pen Moss - Peel Park and Peel Park - Pen Moss. Beneath them, on the line "Goodshaw Hill" are written the observed final bearings, Pen Moss - Goodshaw Hill, Peel Park - Goodshaw Hill. The four bearings must be in cyclic order. The difference between the left-hand pair is the angle at Pen Moss; between the right-hand pair, the angle at Peel Park; and between the pair on the line "Goodshaw Hill", the angle at Goodshaw Hill. These angles are written opposite their respective stations and the sum checked to be 180°00'00".

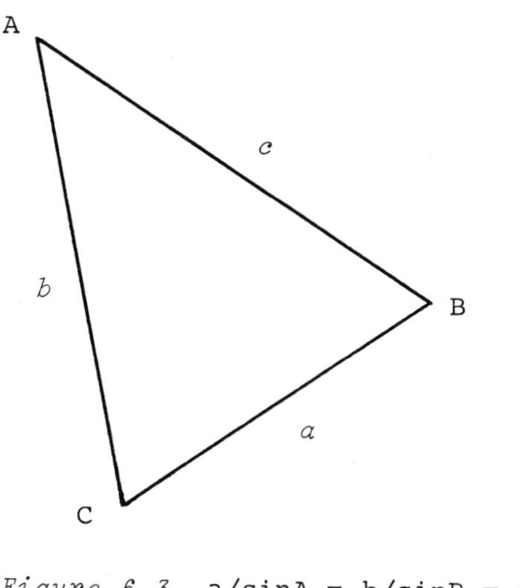

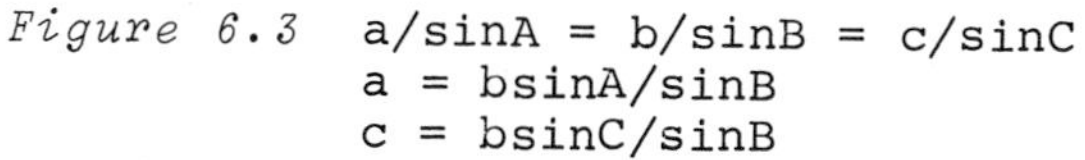

Figure 6.3 $a/\sin A = b/\sin B = c/\sin C$
$a = b\sin A/\sin B$
$c = b\sin C/\sin B$

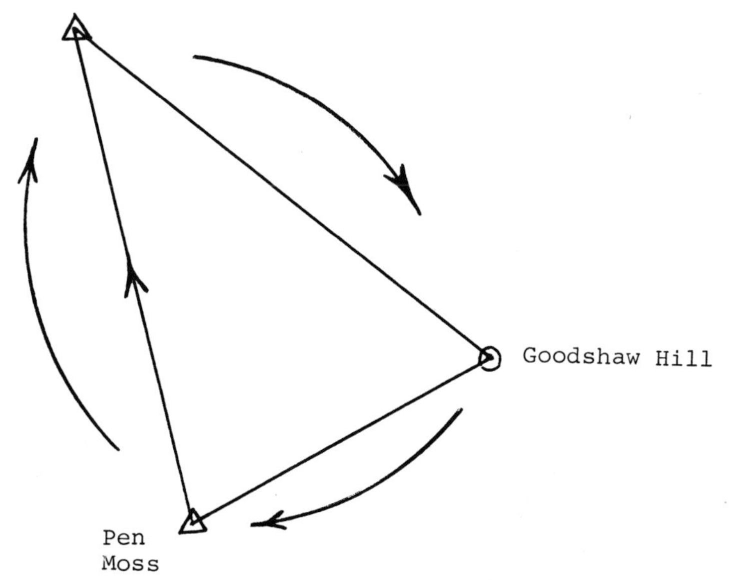

Figure 6.4

Example

GOODSHAW HILL

Pen Moss - Peel Park

+ 16315.030	+ 3374.670	3.4850681
+ 19370.430	+ 2235.370	3.0566381
+ 3055.400	- 1139.300	0.4284300
339.33.02	3260.899	9.9717307
		3.5133374

Peel Park	23.50.24	9.6065791		
Pen Moss - Peel Park		3.5133374	339.33.02	159.33.02
Goodshaw Hill	69.26.03	0.0285993	66.16.35	135.42.38
Pen Moss	86.43.33	9.9992905		
	180.00.00			

3.1485158	3.1485158	3.5412272	3.5412272	
9.6045771	9.9616569	9.8548047	9.8440317	
2.7530929	3.1101727	3.3960319	3.3852589	
+ 566.360	+ 1288.762	- 2489.040	+ 2428.057	
+ 16315.030	+ 3374.670	+ 19370.430	+ 2235.370	
+ 16881.390	+ 4663.432	+ 16881.390	+ 4663.427	Prov.

GRAPH

Cribden Hill - Goodshaw Hill		354.15.58
Cribden Hill	+14040.100	+4948.760
	2841.290	-285.296
	3.4535156	+4663.464
	9.0017799	0.032
	2.4552955	
Haslingden Moor - Goodshaw Hill		35.45.57
Haslingden Moor	+13674.350	+2353.420
	3207.040	+2310.084
	3.5061044	+4663.504
	9.8575234	0.072
	3.3636278	
Great Hameldon - Goodshaw Hill		173.08.29
Great Hameldon	+18932.200	+4416.740
	2050.810	+246.672
	3.3119254	+4663.412
	9.0801949	
	2.3921203	0.020
Hameldon Eccentric - Goodshaw Hill		172.04.37
Hameldon Eccentric	+18911.960	+4380.830
	2030.570	+282.598
	3.3076180	+4663.428
	9.1435507	
	2.4511687	0.004

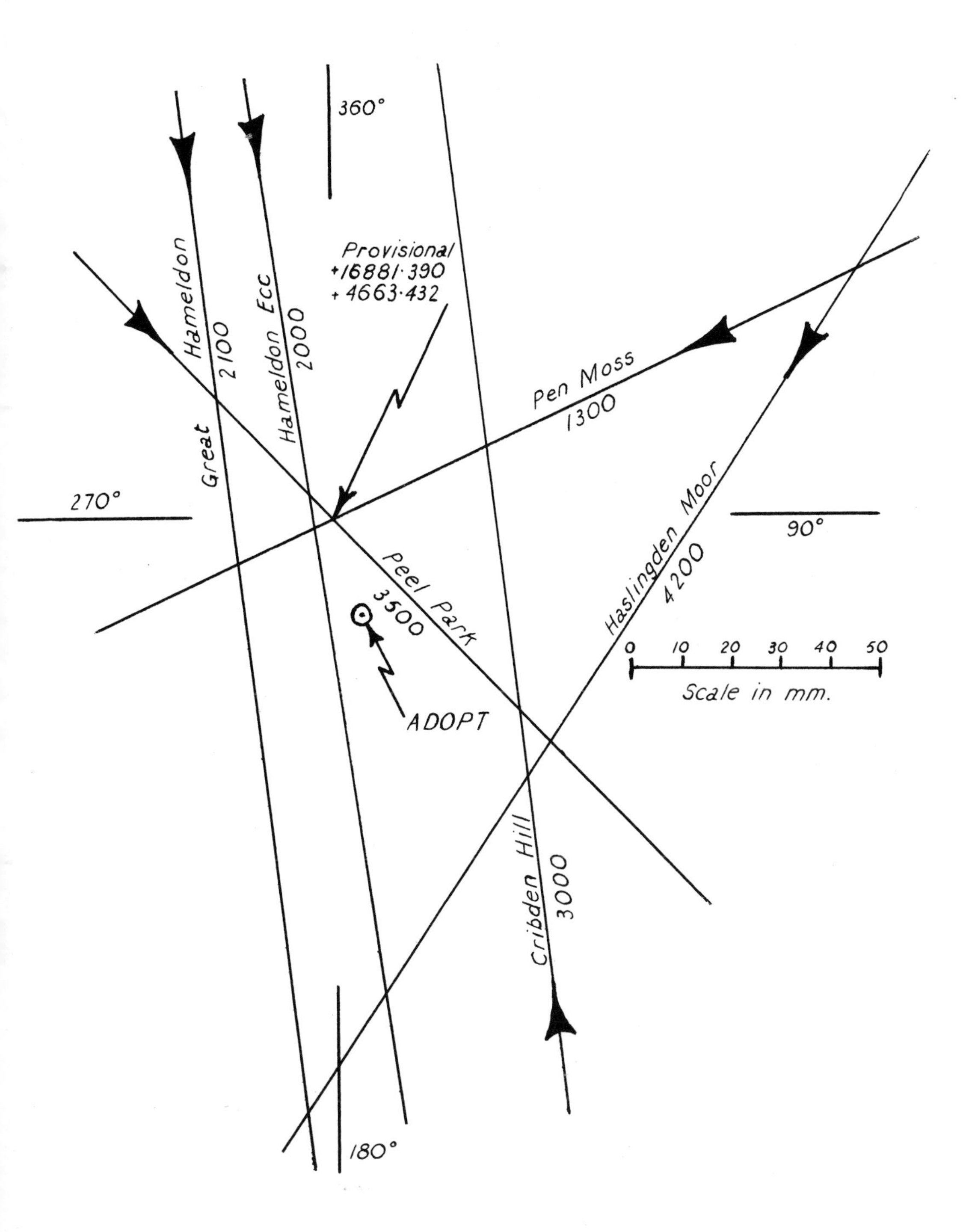

GOODSHAW HILL
360°
Provisional
+16881·390
+4663·432
Great Hameldon
2100
Hameldon Ecc
2000
Pen Moss
1300
270°
90°
Haslingden Moor
4200
Peel Park
3500
0
10
20
30
40
50
Scale in mm.
ADOPT
Cribden Hill
3000
180°
GOODSHAW HILL +16881·371 +4663·437

Figure 6.5

The next paragraph refers to the logarithmic calculations.

Opposite their names, the sines of the angles at Peel Park and Pen Moss are written, followed by the inverse of the sine of the angles at Goodshaw Hill (obtained by subtracting the sine from 10 directly from the tables). The sum of the upper three figures gives the length Pen Moss to Goodshaw Hill and is twice written below on the left-hand side of the sheet; the sum of the lower three is the length, Peel Park to Goodshaw Hill, written similarly but on the right. Beneath them, in the order cosine first and sine second, are the bearings Pen Moss to Goodshaw Hill on the left, Peel Park to Goodshaw Hill on the right. The sums of these pairs of logarithms give the respective latitudes and departures of Pen Moss to Goodshaw Hill, Peel Park to Goodshaw Hill and the signs shown in Shortrede's Tables (cos first, sin second) are applied to the antilogarithms which are written underneath.

The coordinates of the known stations are applied to these to give the provisional coordinates of Goodshaw Hill which are thus duplicated to check the calculation.

The strike of rays from all the other stations, relative to these provisional coordinates, is then calculated. The difference between the Y or X coordinates, whichever be the larger, of the known point and the provisional point, multiplied by the tangent or cotangent of the bearing, whichever be the smaller (it is always in the second column in Shortrede's Tables), gives the X or Y coordinate difference. This is applied to the coordinate of the known station and may result in a larger or smaller value than the respective provisional coordinate. The difference is noted and identified by enclosing it in a box.

The information is plotted to a suitable scale on graph paper by marking on the appropriate axis the distance by which a given ray misses the "provisional" position and setting out its bearing by protractor.

Observations which appear to be faulty and those taken in one direction only should be given less weight and the adopted position may be assumed to be such that its distance from a fixing ray is proportional to the length of that ray. The final coordinates may be found by inspection.

The accuracy of the framework may be inferred from the diagrams plotted for each station but, of course, may not exceed the accuracy of the base line. Ordnance Survey Triangulation Stations may be taken to give normally an accuracy of 1/100000 to 1/200000. Base lines measured with a steel band, however carefully, cannot be guaranteed to better accuracy than 1 part in 20000.

7 Geodimeter Surveys

The geodimeter may be used in conjunction with the theodolite for triangulation, trilateration, traverse surveys, trigonometrical levelling or frameworks combining all of these. The calculations described here are an extension of those set out in the chapters dealing with each method of surveying.

In order to match the accuracy of the geodimeter, horizontal and vertical angles should be measured with a theodolite reading to one second of arc.

Calculations

(a) Conversion of geodimeter slope distance to the respective horizontal and vertical distances

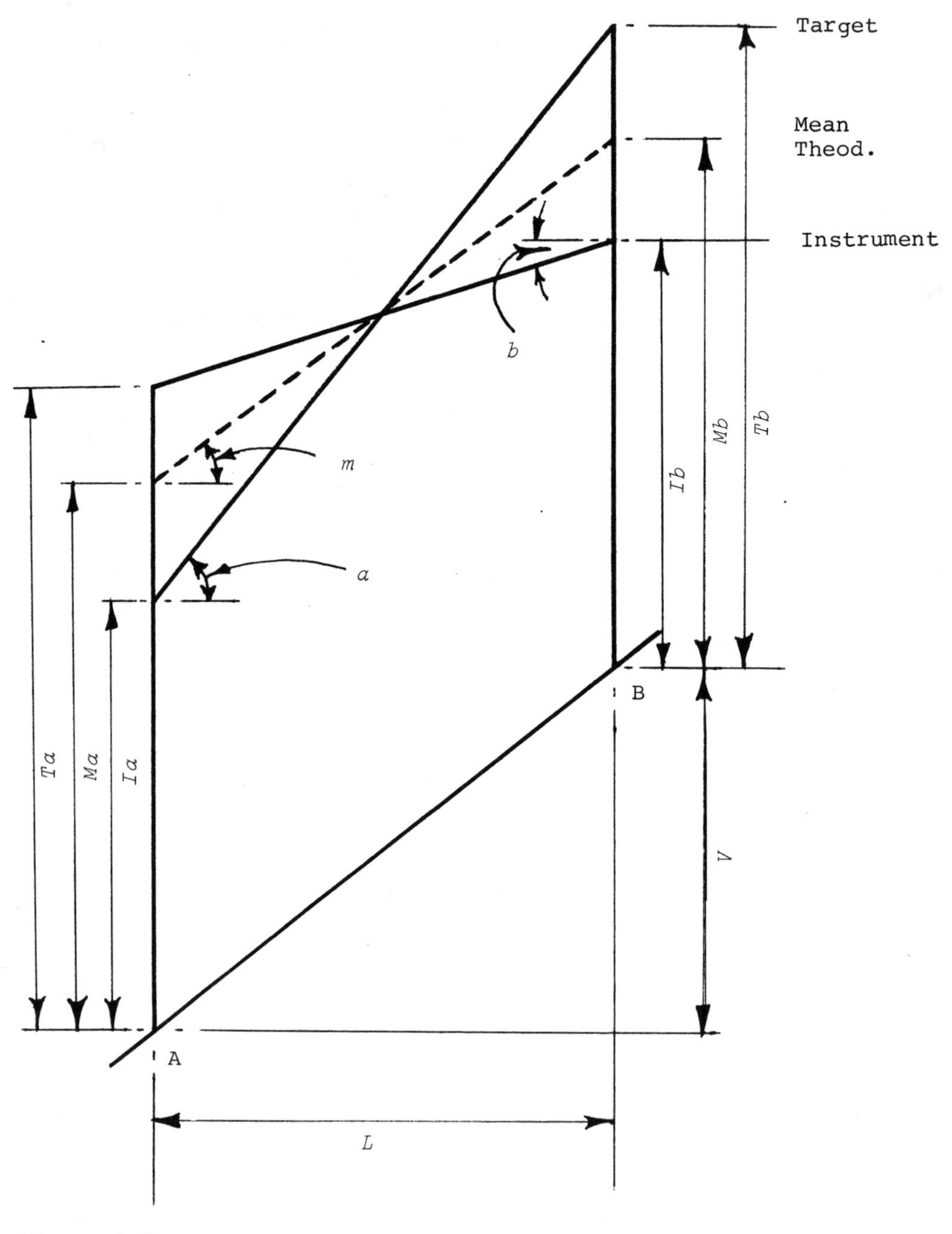

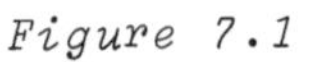

Figure 7.1

Let $\frac{1}{2}(a + b) = m$ $\qquad \frac{1}{2}(I + T) = M$

Si = slope distance between mean theodolite points.

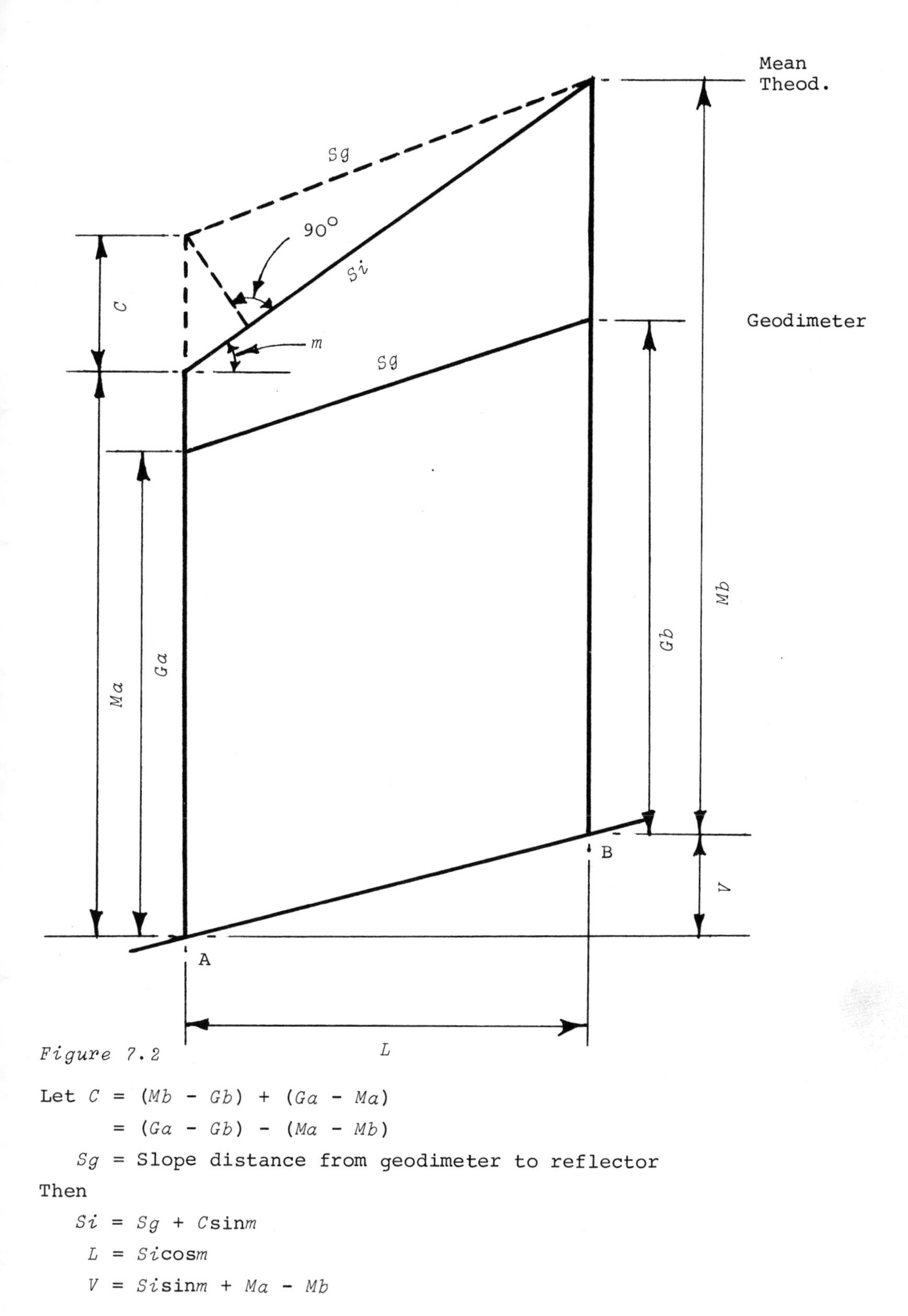

Figure 7.2

Let $C = (Mb - Gb) + (Ga - Ma)$

$= (Ga - Gb) - (Ma - Mb)$

Sg = Slope distance from geodimeter to reflector

Then

$Si = Sg + C\sin m$

$L = Si\cos m$

$V = Si\sin m + Ma - Mb$

The above statements are not mathematically exact but are sufficiently accurate for practical purposes.

Examples of Conversion of Slope Lengths

JARROW	- 0°00'42"	+ 1.520	+ 0.320	+ 1.020	2283.949
HOWDON	- 0.00.11	- 0.240	- 1.580	- 1.500	
	2)0.00.31	+ 1.280	- 1.260	- 0.480	
	- 0.00.15	+ 0.640	- 0.630	+ 0.010	5.8616661
		+ 0.010		- 0.490	9.6901961
				- nil	5.5518622
				2283.949	
				2283.949	

		3.3586864
	- 0.00.15	0.0000000
	2283.949	3.3586864
		5.8616661
		9.2203525
OD		
6		
6	2283.949	- 0.166
2)12(6	- 0.002	+ 0.010
Jarrow Howdon	2283.947	- 0.156

Jarrow	- 0.00.42	+ 1.520	+	+ 1.020	2283.949
Howdon	-	-	- 1.580	- 1.500	
2280 app.	+ 32			- 0.480	
	- 0.00.10	+ 1.520	- 1.580	- 0.060	5.6855749
		- 0.060		- 0.420	9.6232493
				- nil	5.3088242
				2283.949	
				2283.949	

		3.3586864
	- 0.00.10	0.0000000
	2283.949	3.3586864
		5.6855749
		9.0442613
OD		
6		
6	2283.949	- 0.111
2)12(6	- 0.002	- 0.060
Jarrow Howdon	2283.947	-00.171

JARROW a Ia Ta Ga Sg
HOWDON b Ib Tb Gb

2) $a + b$ $Ia - Ib$ $Ta - Tb$ $Ga - Gb$

m $Ma - Mb$ sin$m$

$Ma - Mb$ C C

corr. corr.

Sg

Si

logs

alt. length
m cos$m$
Si Si
sin$m$
Vert. dist.

	Alt. length Corr. to sea	Vert. dist. $Ma - Mb$
Jarrow Howdon	L	V

JARROW a Ia Ga Sg
HOWDON Tb Gb

Curv. corr. $Ga - Gb$

m $Ia - Tb$ sin$m$

$+ Ia - Tb$ C C

corr. corr.

Sg

Si

All as above.

DAM	+ 28.30.15	+ 1.480	+ 1.900	+ 1.250	704.766
CRUACHAN	- 28.28.01	- 1.420	- 1.570	- 1.400	
	2)56.58.16	+ 0.060	+ 0.330	- 0.150	
	+ 28.29.08	+ 0.030	+ 0.165	+ 0.195	9.6784612
			+ 0.195	- 0.345	9.5378191
				- 0.164	9.2162803
				704.766	
				704.602	
		2.7919019			
	+ 28.29.08	9.9439579			
	704.602	2.8479440			
		9.6784612			
OD		2.5264052			
435					
770	619.301	+ 336.051			
2)1205(602	- 0.058	+ 0.195			
Dam-Cruachan	619.243	+ 336.246			

Note: To obtain the correct sign and quantity for the correction *Sg* to *Si*, subtract the mean theodolite heights from the geodimeter heights.

Tables for curvature/refraction corrections and length-altitude corrections are given in the section on trigonometrical levelling.

(b) Traverse and triangulation computations

The horizontal sea level distances and the relative heights of stations may be adjusted for errors in a purely traverse network by the method indicated in the section on traverse calculations.

Where the framework is a development of the triangulation system, an extension of the plotting system described in the section on triangulation may be found to be the simplest approach.

Two cases must be considered. One where the vertical angles and the distances are small, up to about four degrees and 1250 m (4000 ft); the other where either limit is exceeded.

In the first case, errors in the vertical angles will not affect sensibly the accuracy of the horizontal distances derived from the geodimeter measurements. Consequently, they may be plotted directly on the plan view of the station.

In the example, the various positions for Ash Filter given by the bearings and distances observed from other stations are plotted to some suitable scale on graph paper. The respective bearings of the rays through each of these points are set out with a protractor and the arc denoting the observed distance from these stations to Ash Filter is drawn as a straight line at right angles to the angular rays. The final coordinates of the station may then be deduced by inspection.

In the second case, errors in vertical angles will give discrepancies not only in the heights obtained for the station but will also affect the value of the horizontal distances between stations. A convenient way of reconciling the various errors is to plot first the rays striking a station in vertical elevation. This is done by plotting the various derived heights on graph paper and setting out with a protractor the respective vertical angles.

By inspection, the best value for the reduced level may be adopted. Then, by plotting lines at right angles to the observed rays, the change in the horizontal distances given by the adjustment of the vertical angles may be scaled along the line showing the adopted height. This change, in conjunction with the coordinates and bearings obtained from the horizontal rays, may be plotted on the plan view.

Example

Given:			
London Airport - Ash Filter		169°16'03"	2043.939
Mary Valve - " "		133.32.21	332.824
Jobs - " "		234.44.52	966.398
Mary East - " "		111.14.31	157.580
Jobs	+ 1096.979	+ 2657.010	
London Airport	+ 2547.371	+ 1487.188	
Mary East	+ 596.284	+ 1720.959	
Mary Valve	+ 768.462	+ 1626.573	

ASH FILTER

L.A.P.		Mary East	
	3.3028050		1.7565778
169.16.03	9.9923358	111.1431	9.5590767
2043.945	3.3104692	157.580	2.1975011
	9.2700355		9.9694433
	2.5805047		2.1669444
- 2008.191	+ 380.632	- 57.092	+ 146.874
+ 2547.371	+ 1487.188	+ 596.284	+ 1720.959
+ 539.180	+ 1867.820	+ 539.192	+ 1867.833
Provisional coordinates		0.012	0.013

Mary Valve		Jobs	
	2.3603394		2.7464649
133.32.21	9.8381248	234.44.52	9.7613089
332.824	2.5222146	966.398	2.9851560
	9.8602803		9.9120196
	2.3824949		2.8971756
- 229.266	+ 241.265	- 557.783	- 789.179
+ 768.462	+ 1626.573	+ 1096.979	+ 2657.010
+ 539.196	+ 1867.838	+ 539.196	+ 1867.831
0.016	0.018	0.016	0.011

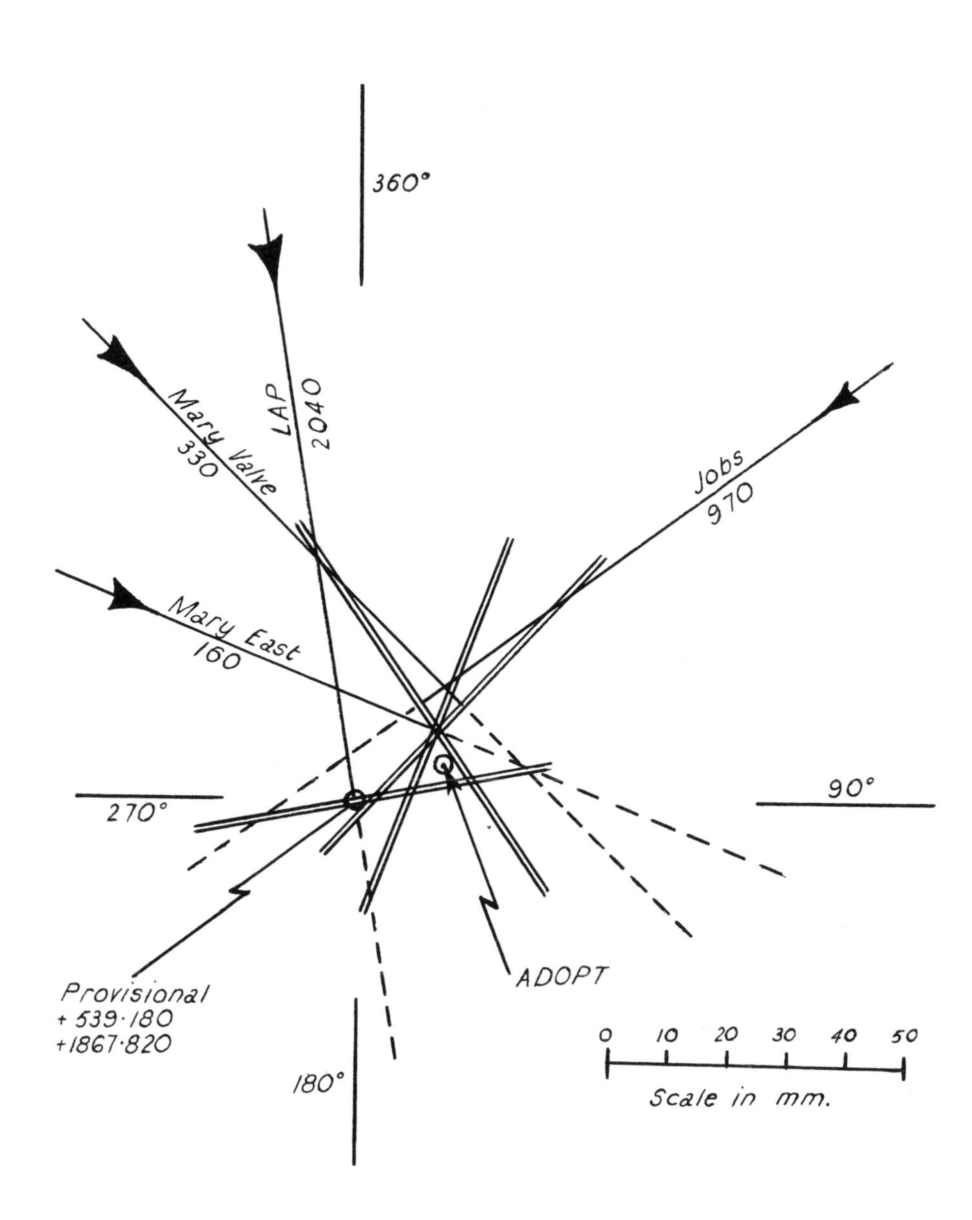

ASH FILTER + 539·186 +1867·835

Figure 7.3

Example

Given:

(i) Knoll R.L. from South Tree 339.750 OD

(i) Knoll R.L. from	South Tree	339.750	OD
" " "	Dam	339.690	
" " "	North Loch	339.675	
" " "	Hill Road	339.655	
" " "	South Loch	339.590	

(ii) Observed bearings and vertical angles, and horizontal distances given by measurement of slope lengths by geodimeter, to Knoll from:

	H		V	
South Tree	H	$256^{\circ}20'31''$	V + $24^{\circ}50'52''$	2149.610
Dam		20.51.00	- 5.01.20	1065.708
North Loch		292.02.40	+ 29.45.17	908.232
Hill Road		292.02.44	+ 30.10.06	255.066
South Loch		268.41.07	+ 20.20.41	1911.391

(iii) Knoll Coordinates derived from:

Hill Road	+ 4071.560	+ 374.880	North Loch	+ 4071.548	+ 374.868
	(Provisional Coordinates)			0.012	0.012
South Tree	+ 4071.540	+ 374.905	South Loch	+ 4071.577	+ 374.871
	0.020	0.025		0.017	0.009
Dam	+ 4071.567	+ 374.897			
	0.007	0.017			

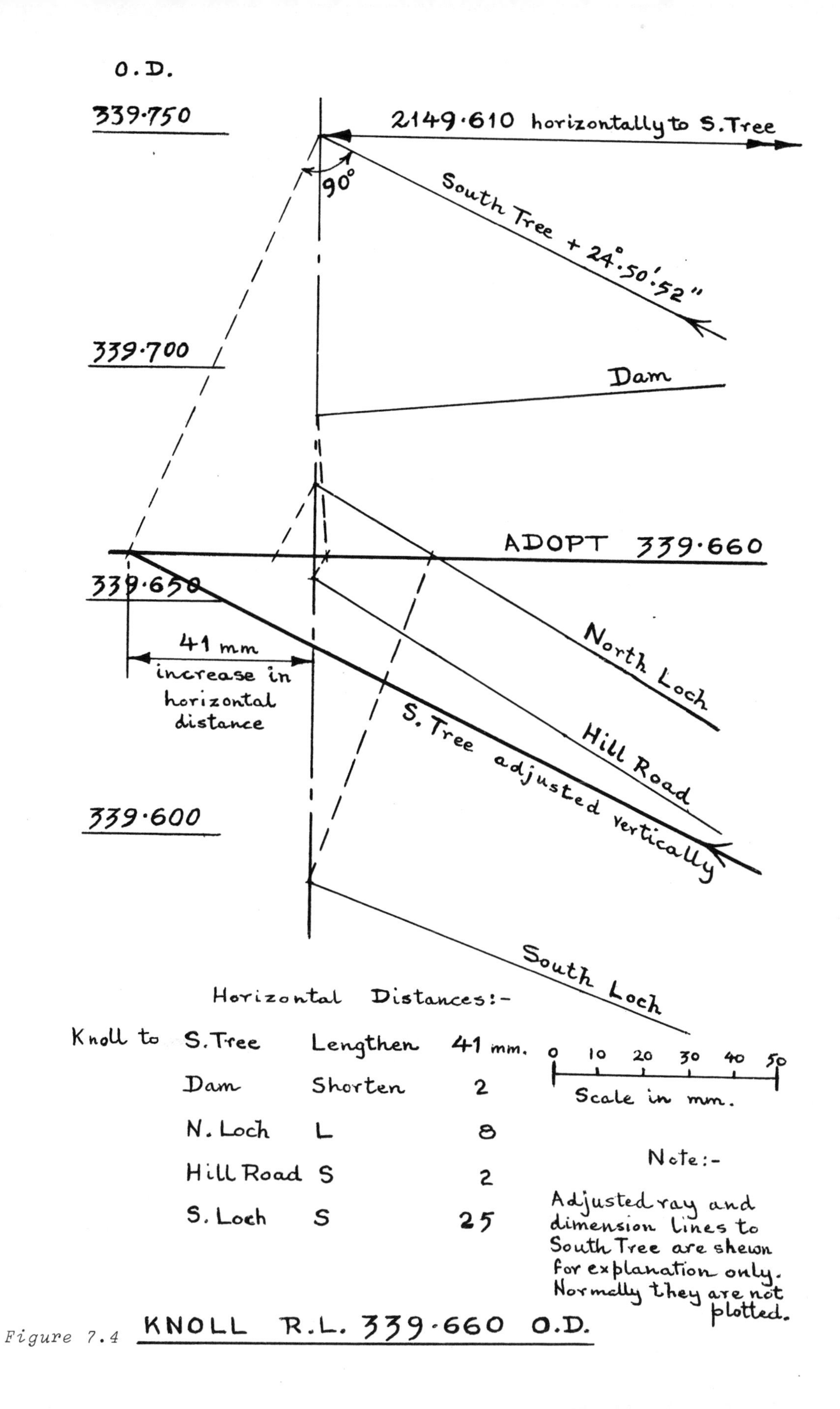
O.D.
339·750
2149·610 horizontally to S.Tree
90°
South Tree + 24°50'52"
339·700
Dam
ADOPT 339·660
339·650
41 mm
increase in horizontal distance
North Loch
Hill Road
S. Tree adjusted vertically
339·600
South Loch
Horizontal Distances:-
Knoll to S.Tree Lengthen 41 mm.
Dam Shorten 2
N. Loch L 8
Hill Road S 2
S. Loch S 25
0 10 20 30 40 50
Scale in mm.
Note:-
Adjusted ray and dimension lines to South Tree are shewn for explanation only. Normally they are not plotted.
KNOLL R.L. 339·660 O.D.

Figure 7.4

KNOLL

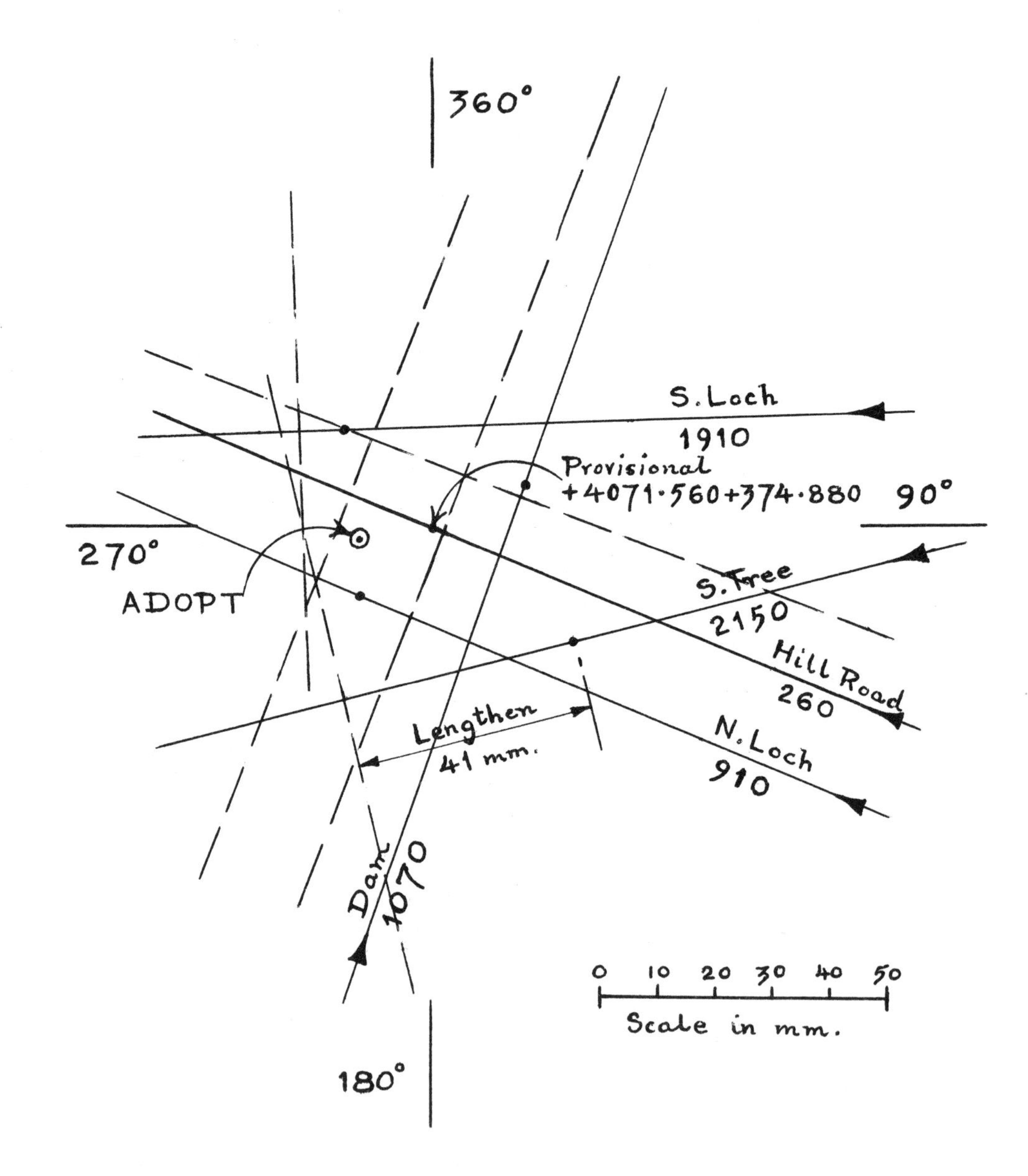

KNOLL + 4071·558 + 374·867

Figure 7.5

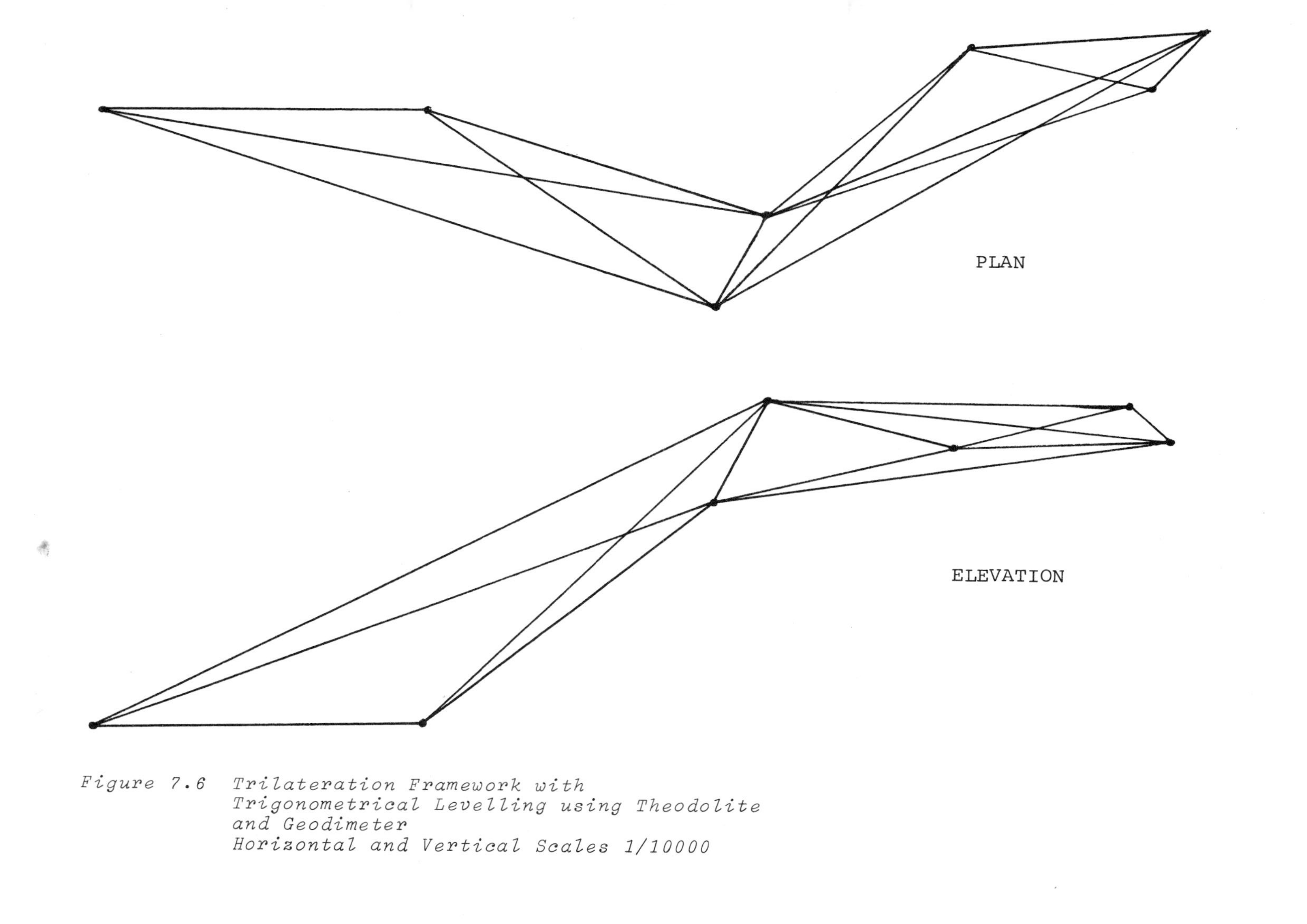

Figure 7.6 Trilateration Framework with Trigonometrical Levelling using Theodolite and Geodimeter Horizontal and Vertical Scales 1/10000

8 The Three Point Problem

The example given below sets out a method for deriving the coordinates of a new station, Pillbox, relative to three Ordnance Survey Stations and checking them against a fourth O.S. Station. Angles are observed at the new station only.

AT PILLBOX

F.B. Readings

London Airport	$336^{\circ}14'01''$
O.S. 98B	27.28.30
O.S. 98A	64.01.44
O.S. 87A	209.30.51

Given, coordinates:

	Y	*X*
London Airport	+ 2547.371	+ 1487.188
O.S. 98B	+ 923.256	+ 2594.876
O.S. 98A	+ 917.123	+ 2602.724
O.S. 87A	+ 888.411	+ 2583.248

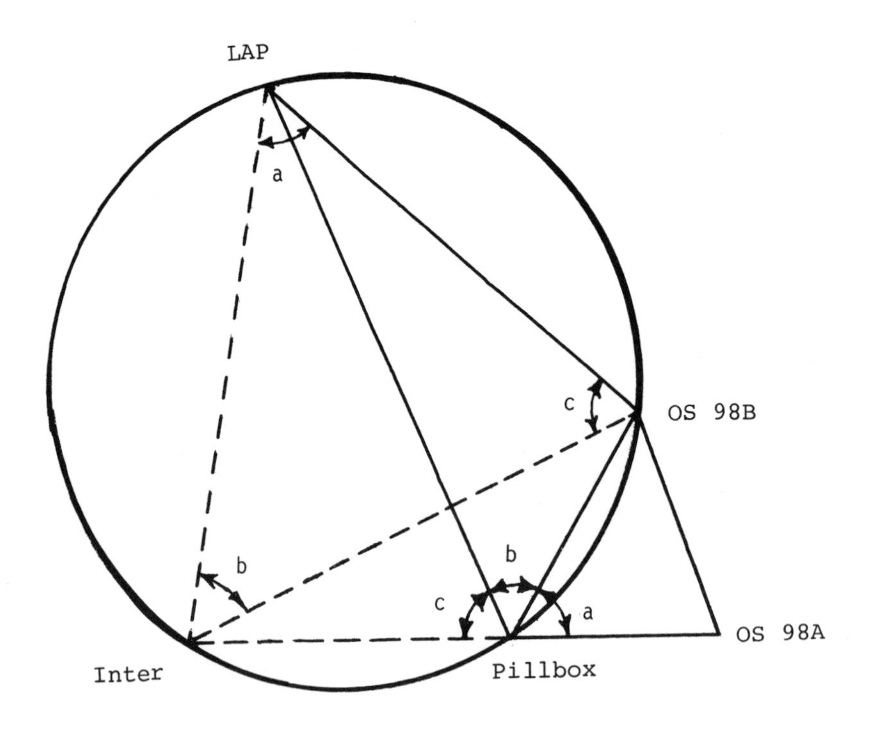

Figure 8.1

O.S. 98A	64.01.44	
O.S. 98B	27.28.30	
		Angle *a*

O.S. 98B	27.28.30	
L.A.P.	336.14.01	
	51.14.29	
		Angle *b*

L.A.P.	336.14.01	
O.S. 98A	244.01.44	reciprocal bearing
	92.12.17	
		Angle *c*

INTER

L.A.P. - O.S. 98B

+ 2547.371	+ 1487.188	3.2106168
+ 923.256	+ 2594.876	3.0444176
- 1624.115	+ 1107.688	0.1661992
		9.9170576
145.42.18		3.2935592
1965.890		

O.S. 98B	92.12.17	9.9996784		
L.A.P. - O.S. 98B		3.2935592	145.42.18	325.42.18
Inter	51.14.29	0.1080221	182.15.32	233.30.01
L.A.P.	36.33.14	9.7749391		
	180.00.00			

3.4012597	3.4012597	3.1765204	3.1765204
9.9996624	8.5956597	9.7743848	9.9051803
3.4009221	1.9969194	2.9509052	3.0817007
- 2517.225	- 99.293	- 893.110	- 1206.981
+ 2547.371	+ 1487.188	+ 923.256	+ 2594.876
+ 30.146	+ 1387.895	+ 30.146	+ 1387.895

Inter - O.S. 98A

+ 30.146	+ 1387.895	2.9479123
+ 917.123	+ 2602.724	3.0845152
+ 886.977	+ 1214.829	9.8633971
		9.9072170
53.51.57		3.1772982
1504.175		

Inter - O.S. 98A = Pillbox - O.S. 98A = $53^{o}51'57$ Grid bearing

Therefore: At Pillbox **- 10.09.47**

London Airport	336.14.01		326.04.14
O.S. 98B	27.28.30		17.18.43
O.S. 98A	64.01.44	(-10.09.47)	53.51.57
O.S. 87A	209.30.51		199.21.04

PILLBOX				
O.S. 98B	128.23.35	9.8941879		
L.A.P. - O.S. 98B		3.2935592	145.42.18	325.42.18
Pillbox	51.14.29	0.1080221	146.04.14	197.18.43
L.A.P.	0.21.56	7.8048278		
	180.00.00			

3.2957692	3.2957692	1.2064091	1.2064091
9.9189345	9.7467677	9.9798664	9.4735948
3.2147037	3.0425369	1.1862755	0.6800039
- 1639.471	+ 1102.902	- 15.356	- 4.786
+ 2547.371	+ 1487.188	+ 923.256	+ 2594.876
+ 907.900	+ 2590.090	+ 907.900	+ 2590.090

Provisional Coordinates

O.S. 98A	233.51.57	O.S. 87A	19.21.04
+ 917.123	+ 2602.724	+ 888.411	+ 2583.248
- 9.224	12.634	19.489	+ 6.844
+ 907.899	1.1015409	1.2897896	+ 2590.092
	9.8633980	9.5455505	
	0.9649389	0.8353401	
0.001			0.002

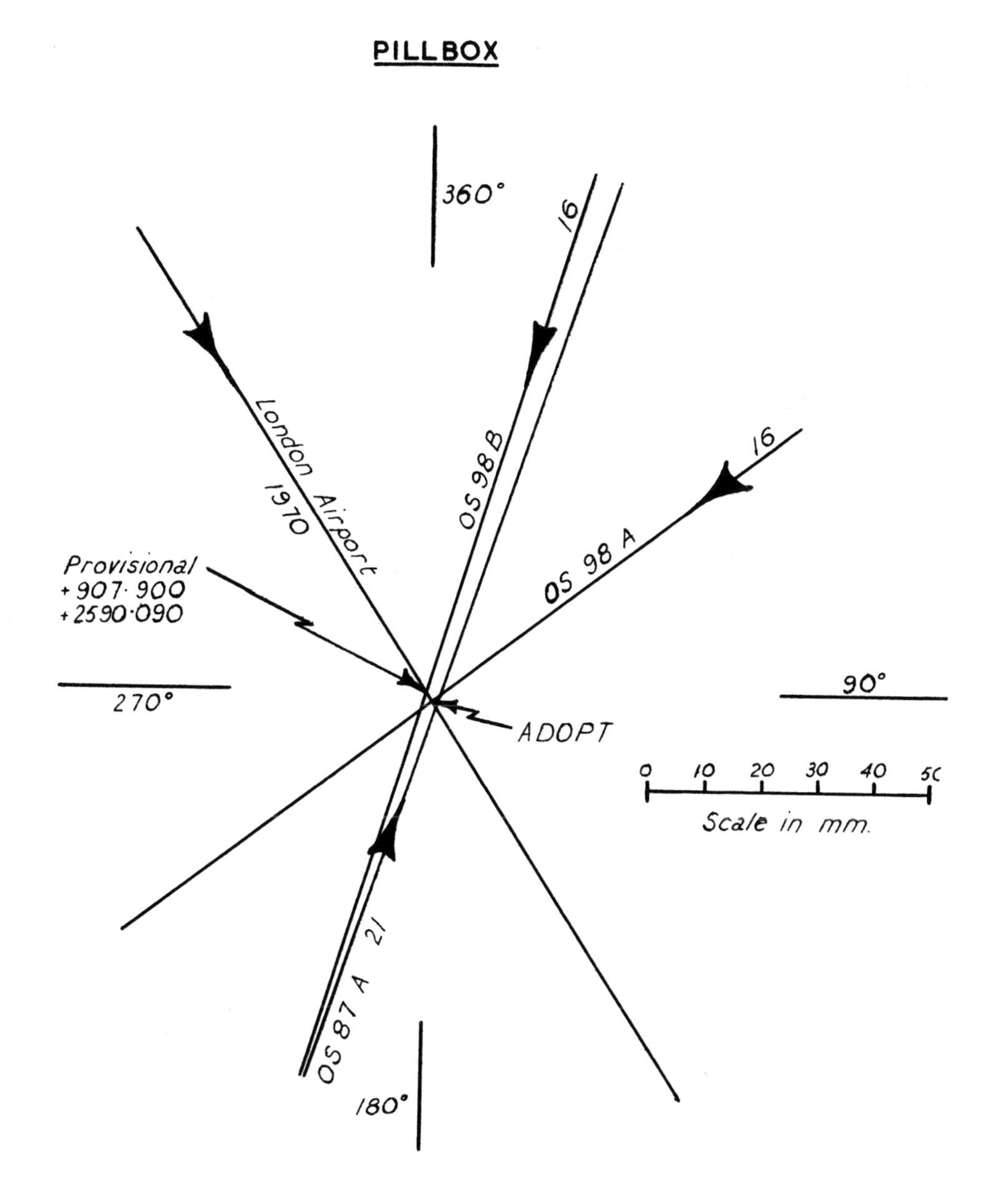

PILLBOX +907·902 +2590·092

Figure 8.2

9 Tacheometric Surveys

The tacheometric survey method described in this chapter is accurate to about one part in 500. It is intended for the preparation of plans showing local detail to a scale of 1/250 or 1/500 and is based on the use of a standard theodolite reading to 20 seconds of arc, in conjunction with a standard staff.

Where a greater accuracy is required (up to say 1/5000) the system may be refined by steadying the staff with stays, estimating the staff readings to 0.5 mm, observing angles on two faces, and using a more accurate set of tacheometric survey tables. In such cases, should there be many points to observe, it is worth considering some other method of survey.

The greatest danger is the strong possibility of mistakes rather than the size of errors. The plotting of each point is an open traverse in three dimensions, each based on several independent readings which may be taken wrongly whilst observing, calculating or plotting. Important positions should be fixed more than once from separate instrument stations or checked by taping and levelling.

For speed and accuracy, two surveyors should work together in all the operations. The number of staffmen depends upon the volume of work to be done and the speed of the observer. Where the maximum of half a dozen chainmen is used, a third surveyor should be available to marshall them.

Observations

The height of collimation of the theodolite is obtained by observing or by setting the instrument upon a station of known reduced level.

Normally, staff readings need be taken to 5 mm (0.01 ft) only, horizontal angles to the nearest 10 or 20 minutes, and vertical angles to the nearest minute as the surveys are rarely plotted to any greater accuracy than this.

The staff should be set plumb with a spirit level and, when sighting the staff, the bottom wire should be set always on an exact metre or foot. This enables the surveyor recording the results to see misreadings quickly by comparing the two intercepts before he moves the chainman. It also makes the reduction of distance from the intercept readings faster and less liable to mistakes. It is convenient sometimes to set the instrument level however as this avoids one operation in the calculations. Positions near the theodolite are most conveniently fixed by using a tape.

Calculations

The simplified tacheometric table in Appendix V is based on a horizontal distance of 100 units. By using it in conjunction with a slide rule, instrument readings are converted to lengths and heights.

Plotting

A whole circle protractor should be used to plot the bearings, which should be identified by their station number.

A pin stuck into the centre point facilitates the rapid setting of a scale to mark off the respective distances, the plotted points being identified also by their number. By this method the drawing of rays for each bearing is not necessary and there is less confusion and erasure.

Example of Field Book

At shaft 3 Centre Peg.
Date ...   Time ...
Observer ...   Instrument ...

Collim. 145.94

Stn.	BW	CW	TW	Horiz. Angle	Vert. Angle	100. I	Horiz. Dist.	Vert. Dist.	Staff	Rise/Fall	RL	Remarks
1	1000	2155	3300	250.50	+ 2.12 87.48	230.0	229.5	+ 8.81	- 2.15	- 6.66	139.28	Fence
2	2000	2955	3915	303.30	+ 2.55 87.05	191.5	191.0	+ 9.71	- 2.95	- 6.76	139.18	Fence
3	3000	3680	4350	312.30	LEV	135.0	135.0	-	- 3.68	- 3.68	142.26	Fence Corner
4	2000	2490	2990	325.10	- 2.50 92.50	99.0	99.0	- 4.89	- 2.49	- 7.38	138.56	M.H.Inv.
5	1000	1970	2940	307.00	+ 3.07 86.53	194.0	193.5	+ 10.50	- 1.97	- 8.53	137.41	M.H.Inv.
6	-	1305	-	320.00	LEV	(tape)	16.69	-	- 1.30	- 1.30	144.64	Fence

10 Shaft Plumbing

The most important aspects of shaft plumbing are:

(1) Possible movement of instrument stations above or below ground must not affect the accuracy of the work.

(2) The instrument and target stations should be arranged so that, should any of them move, this may be detected at the time of making the shaft plumbing observations.

(3) Equipment supporting the wires should be of heavy and rigid construction and supported so that it is not affected by vibration due to machinery in or near the shaft.

(4) The winding drums for the plumb-wires should be as large in diameter as possible in order to minimize the tendency for the wires to follow a helical path when suspended in the shaft after having been stored on the drums.

(5) Tensioning weights should be sufficiently heavy to stress the wires to approximately half their ultimate strength.

(6) A meticulous inspection must be carried out over the full height of each wire to ensure that there are no kinks, that nothing is touching it, and that air currents and water splashes are screened.

(7) Where the weights are immersed in buckets of water or oil, these must be mounted so that they do not rock or vibrate.

It is primarily the amount of care taken over these points that determines the accuracy of the shaft plumbing.

Two theodolites reading to a single second of arc are required, one above ground and one below. They should be placed between one and two shaft-diameters from the edge of the shaft, preferably along the centre line of the tunnel. At least two out-stations are desirable with which to orientate the theodolite on the surface; they should be as far away as possible.

The initial plumbings are carried out by co-planing. At the surface, both wires are set by theodolite on to the required bearing and then the instrument below ground is moved into line with the two wires. A centre mark is established on the instrument station and on two more stations set on the same line as far away as possible. The procedure is repeated with both instruments changed to the opposite face and the stations marked with the mean of the two pointings. Both theodolites and both wires are then moved deliberately, re-set and another pair of pointings is taken.

Distances are taped from the instrument to each wire and also between the wires as a check. As a further check the distances between the wires at the surface and underground are compared. Lengths up to about 20 m between stations should be measured with steel tapes; where longer they should be measured by using a steel band.

Subsequent plumbings are done most conveniently by making use of the Weisbach Triangle, examples of which are given below.

First of all, it is of value to consider the following argument leading to the design of a very accurate system of shaft plumbing.

Accepting that surveying is normally more accurate than setting out, consider that a theodolite actually measures only the angle subtended at the instrument by two targets. That is to say, it measures the angle between the stations only when the instrument and the targets coincide truly with them. The operation of centring the instrument and targets is carried out once only whereas the angle may be observed several times. In shafts and tunnels, where these stations are often very close together, a very small lateral setting error at any or all the stations may cause a very large angular error. Such an error will also be introduced should an instrument or target move whilst observing, or again should a station alter its position between plumbings on one day and another. Instrument stations are often unavoidably in vulnerable positions.

The method of co-planing is based entirely on setting out. Targets are set on stations, instruments are set on stations, wires are set by instrument and vice-versa, gravity tends to make the wires plumb only whilst they are not plumb, stations are set by the instrument below ground.

In order then to avoid this as far as possible, permanent targets are established in a safe place sensibly in line with the shaft-plumbing instrument station and the out-stations. The line joining the out-station to the permanent target is then incorporated in the main survey and it is the bearing found from this which is used to control the shaft plumbings. The accurate positioning of the theodolite on the shaft plumbing station now matters no longer, for its orientation and position may always be deduced from the Weisbach triangle formed by the theodolite, the out-station and the permanent target. It is desirable to have two such base lines crossing the instrument position.

The wires may be suspended anywhere within 20 to 30 seconds of arc of the required line, and the bearing of the line joining the wires may be derived from a further Weisbach triangle given by the survey of the wires with the surface theodolite.

Below ground, the instrument is orientated similarly by a Weisbach calculation of the observations on the wires and a further Weisbach Triangle gives the bearing of the line between any two permanent targets in the shaft and/or tunnel which again have been placed sensibly in line with the underground instrument. They are not necessarily in line with the two wires. As these targets need not be sited with a view to occupying them with a theodolite, it is easier to find a safe place for them and to put them farther apart than is normally possible.

Lines may be produced at any angle to this base from any point along its length by placing a theodolite approximately at the intersection of the base and desired line. Observing the base stations and two or more stations on the new line, the system may be related accurately through their interlinked Weisbach Triangles.

A further advantage of this method is that its accuracy may be judged by a study of all the observations and a measure of the probable error may be deduced from it.

Underground, the wires may be observed by either of the following methods. The weights, with vanes attached, may be steadied in drums of water or oil and the middle of the wire observed at the middle of its swing. This swing may be non-existent or it may be very small and very slow indeed, therefore great care has to be taken in the observing. Alternatively, the wire may be watched for several swings against a fine scale fixed close to it and the average position on the scale observed with the theodolite.

An example of the computations for a shaft plumbing consisting of a framework of Weisbach Triangles is given below.

Example

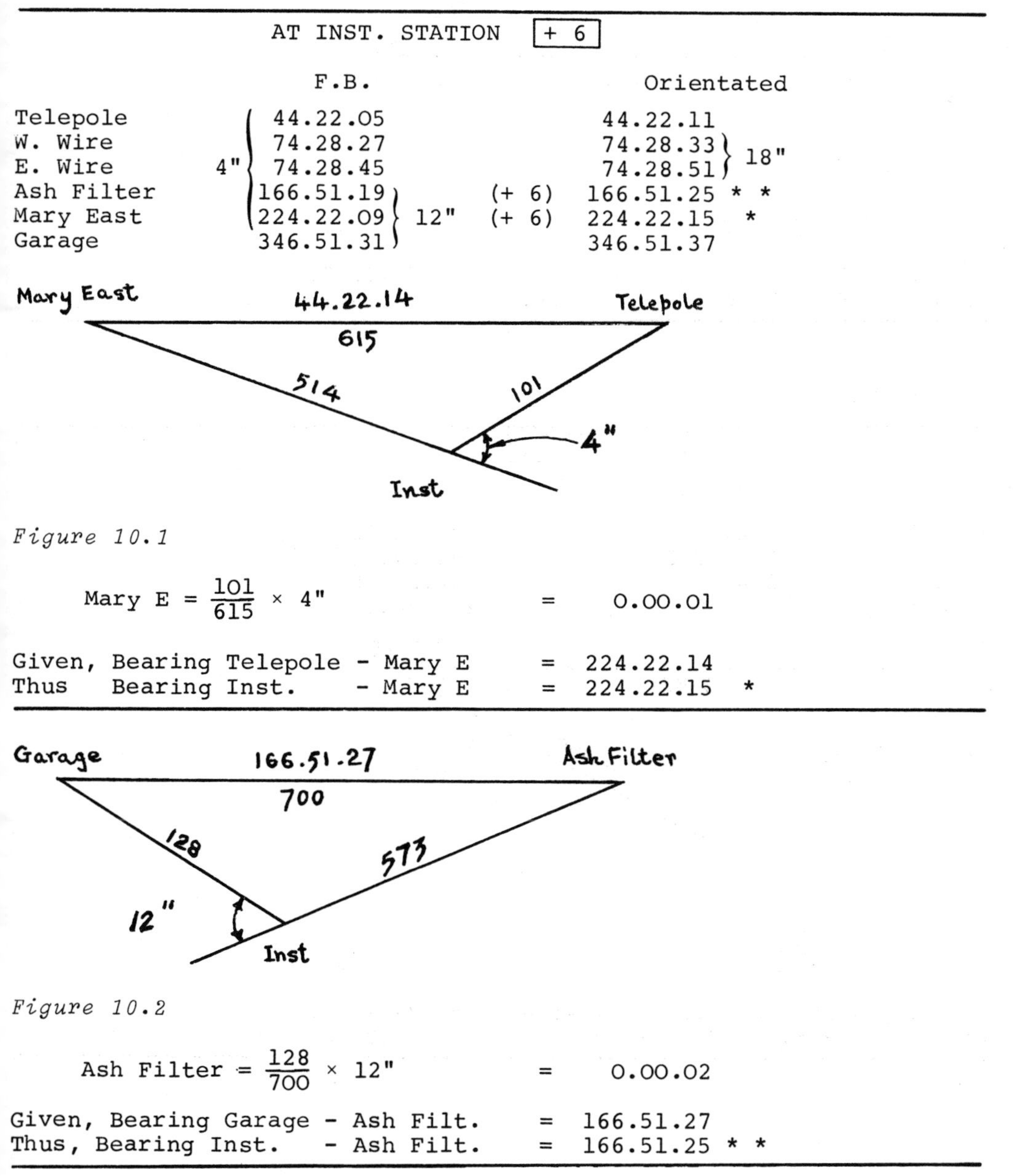

SURFACE

AT INST. STATION [+ 6]

		F.B.			Orientated		
Telepole	4"	44.22.05			44.22.11		
W. Wire		74.28.27			74.28.33	18"	
E. Wire		74.28.45			74.28.51		
Ash Filter		166.51.19	12"	(+ 6)	166.51.25		* *
Mary East		224.22.09		(+ 6)	224.22.15		*
Garage		346.51.31			346.51.37		

Figure 10.1

$$\text{Mary E} = \frac{101}{615} \times 4'' = 0.00.01$$

Given, Bearing Telepole - Mary E = 224.22.14

Thus Bearing Inst. - Mary E = 224.22.15 *

Figure 10.2

$$\text{Ash Filter} = \frac{128}{700} \times 12'' = 0.00.02$$

Given, Bearing Garage - Ash Filt. = 166.51.27

Thus, Bearing Inst. - Ash Filt. = 166.51.25 * *

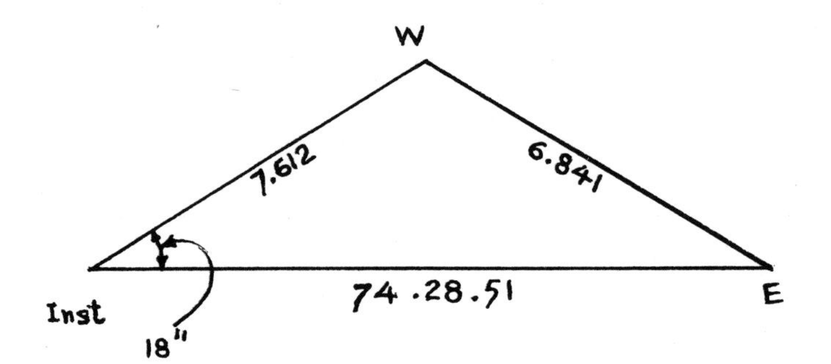

Figure 10.3

$E = \frac{7.612}{6.841} \times 18''$	=	0.00.20
From Obs. Bearing Inst. - E. Wire	=	74.28.51
Thus Brg. W. Wire - E. Wire	=	74.29.11

UNDERGROUND

AT R.S. 3 [- 23"]

		F.B.		Orientated	
R.S. 8		74.28.35		74.28.12	60"
R.S. 2		254.27.35		254.27.12	
E. Wire	61"	254.27.31		254.27.08	
W. Wire		254.28.32	(- 23")	254.28.09	* * *

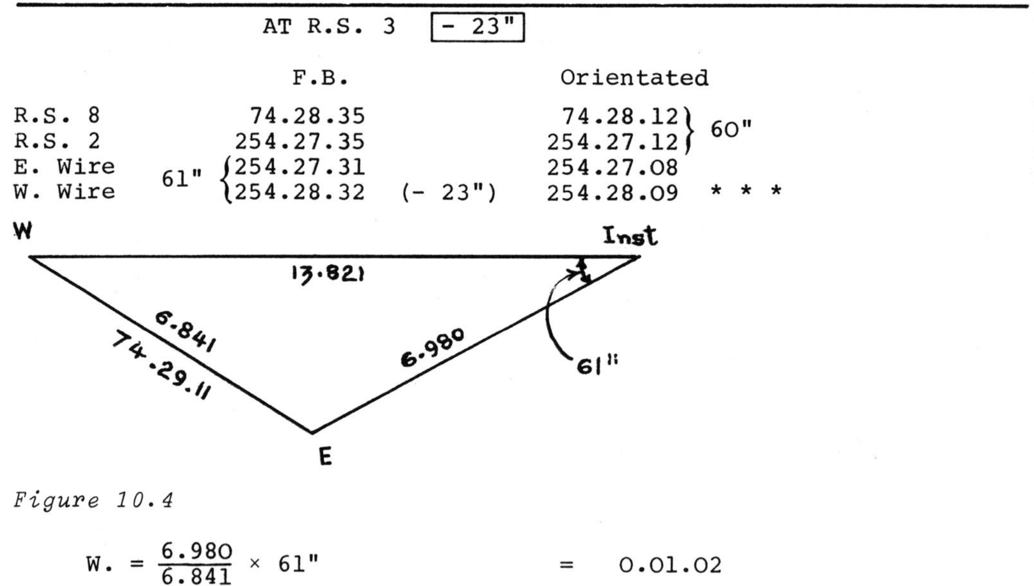

Figure 10.4

$W. = \frac{6.980}{6.841} \times 61''$	=	0.01.02
From Sfce. Brg. W. Wire - E. Wire	=	74.29.11
Thus Brg. W. Wire - Inst.	=	74.28.09 * * *

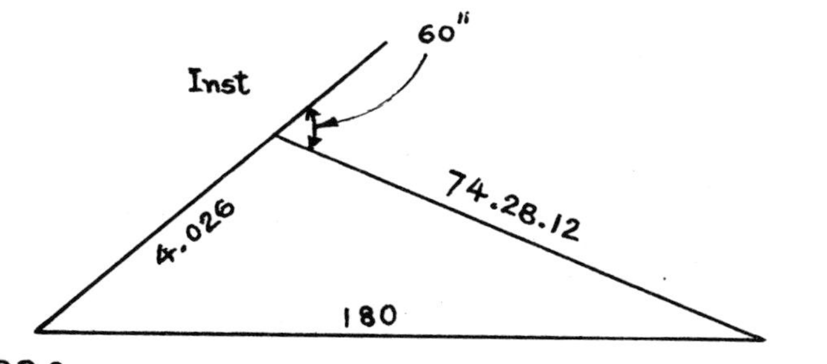

Figure 10.5

	R.S. 8 = $\frac{4.026}{180} \times 60''$	= 0.00.02
From	Obs. Brg. Inst. - R.S. 8	= 74.28.12
Thus	Brg. R.S.2 - R.S. 8	= 74.28.10

Coordinates of the tunnel base line stations may be computed from the bearings and distances obtained and compared with the theoretical tunnel centre lines.

The number of rounds obtained in a full day is normally about eight and the number of days which have to be spent on plumbing a shaft will depend on the agreement between the several values obtained for the underground base-line. Generally three to four complete days' work suffices and a spread of 15 to 20 seconds overall is acceptable.

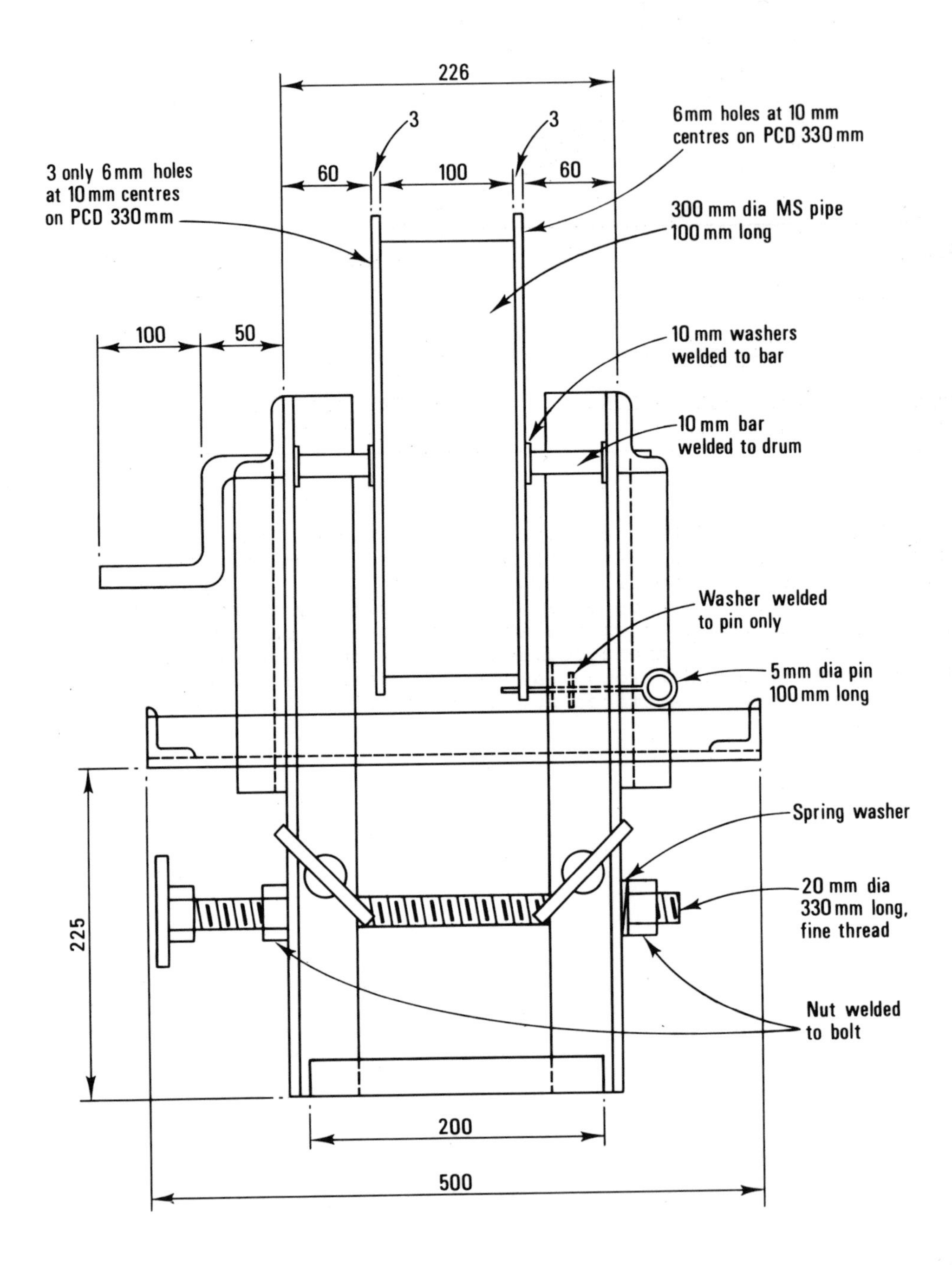

Figure 10.6 Shaft Plumbing - Windlass for Piano Wire Front Elevation

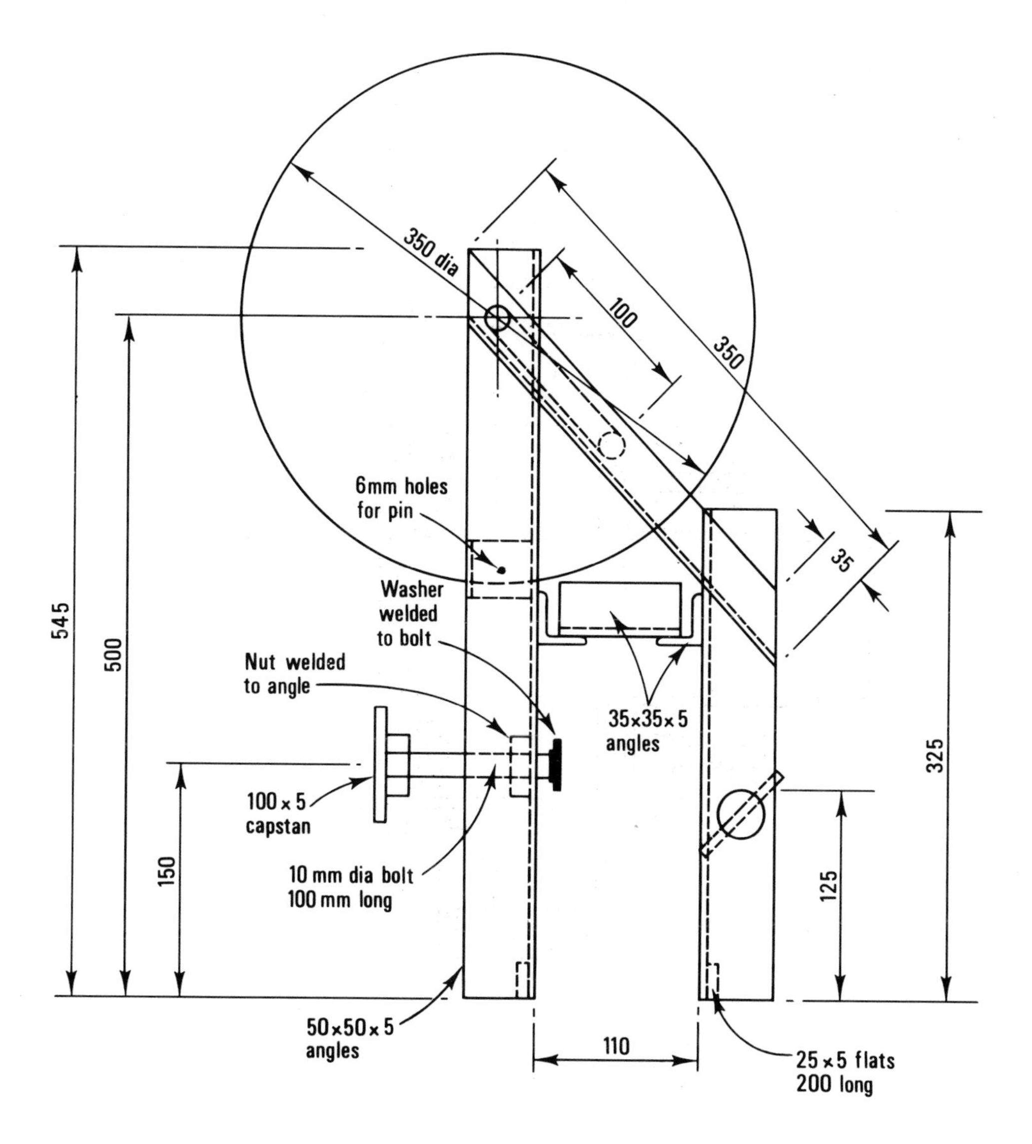

Figure 10.7 Shaft Plumbing - Windlass for Piano Wire Side Elevation

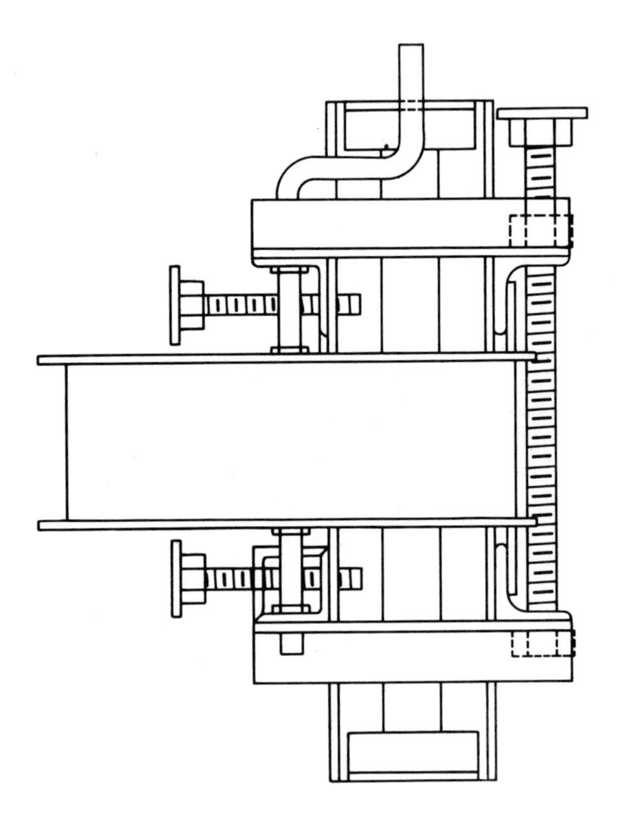

Figure 10.8 Shaft Plumbing - Windlass for Piano Wire Plan

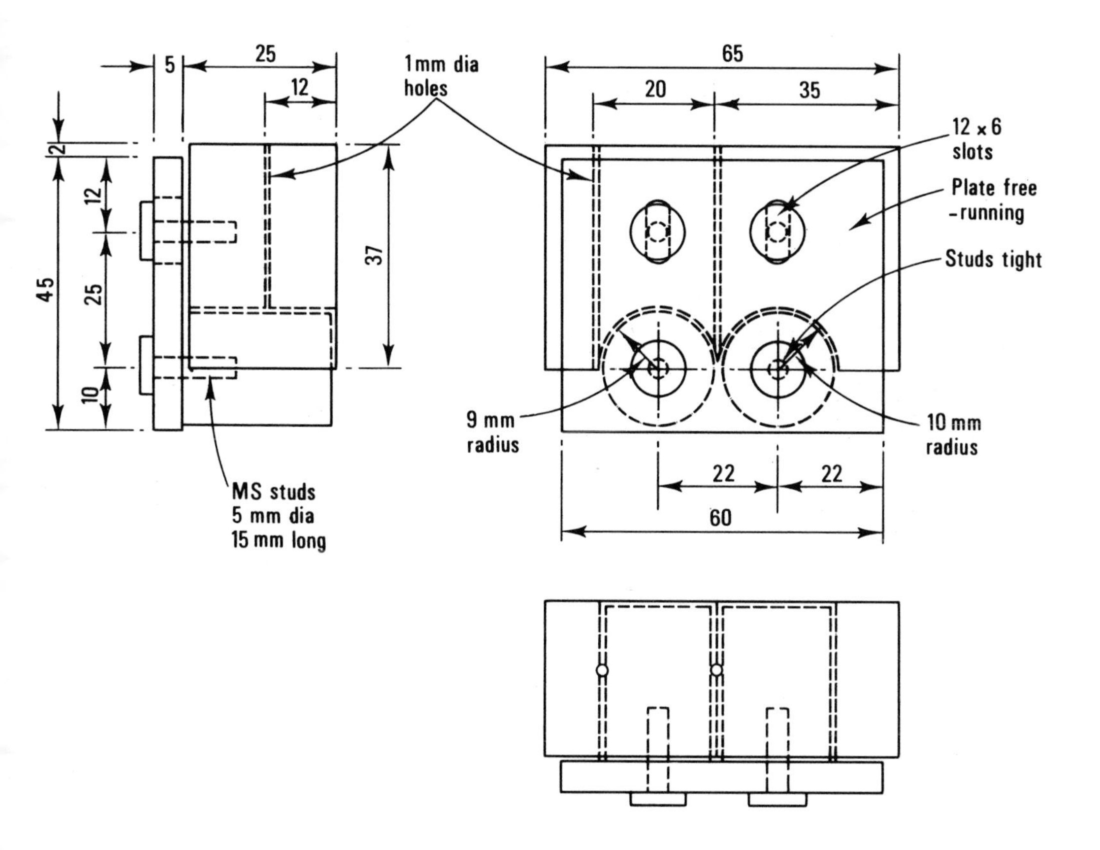

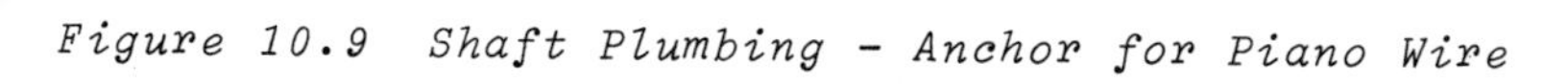
Figure 10.9 Shaft Plumbing - Anchor for Piano Wire

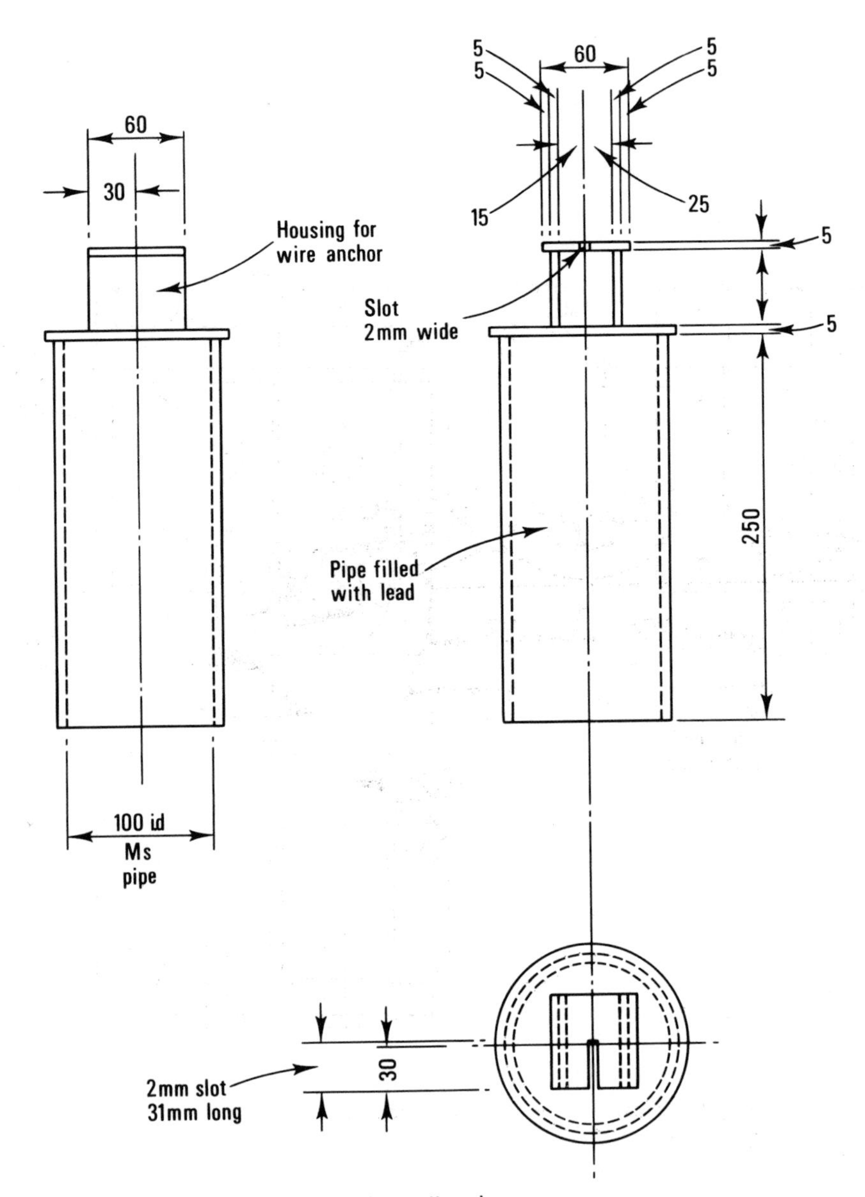

NB. Suitable for 22 SWG, high-tensile wire.

Figure 10.10 Shaft Plumbing - Weight for Plumbing Wire

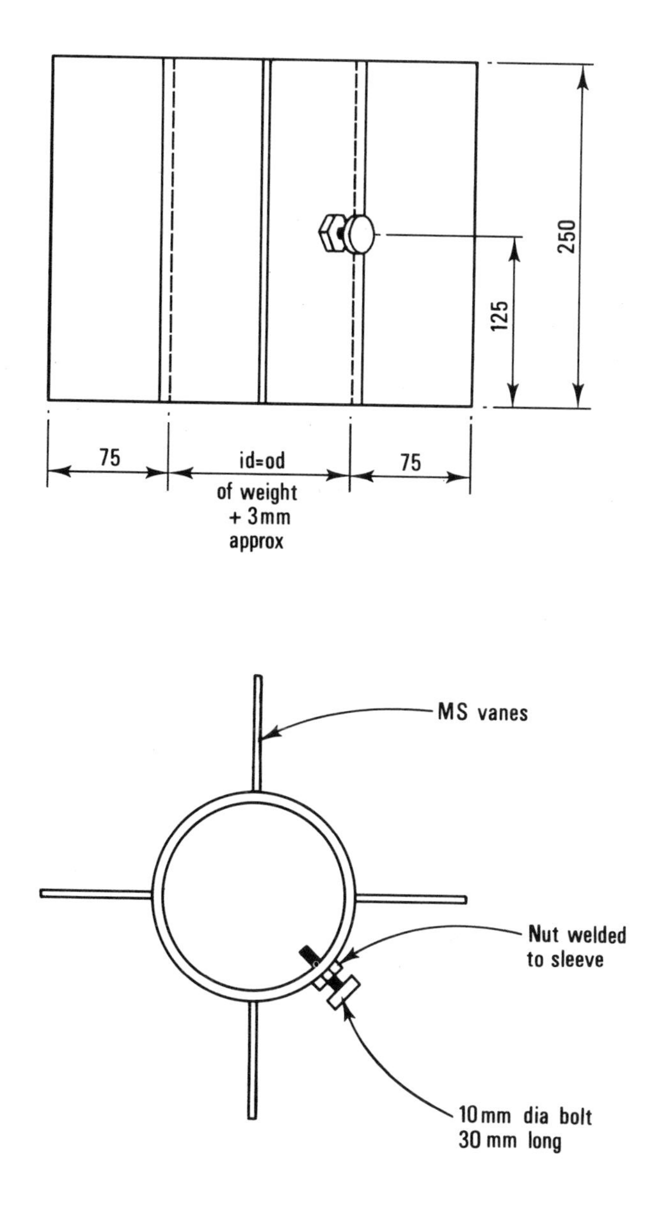

Figure 10.11 Shaft Plumbing - Vanes for Weight

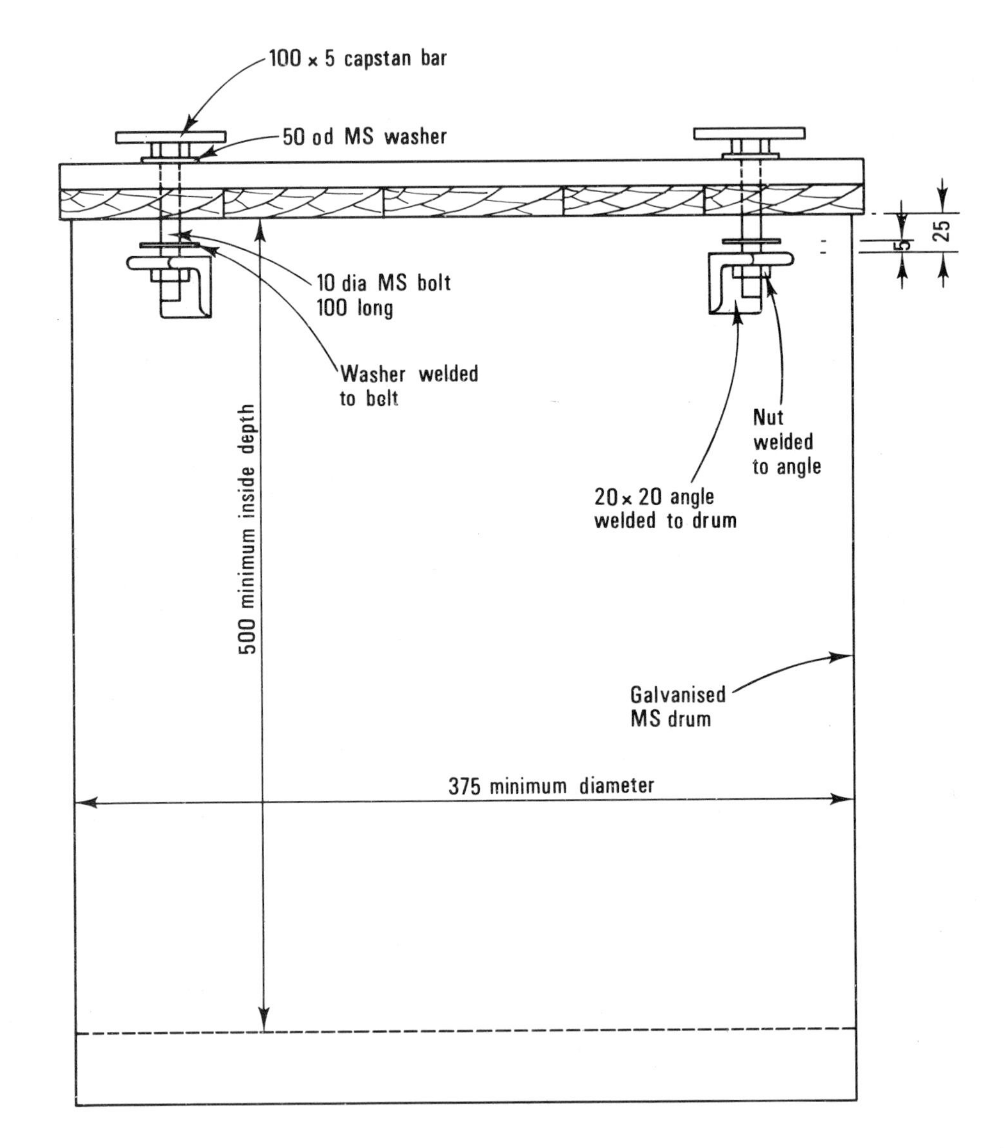

Figure 10.12 Shaft Plumbing - Drum and Lid

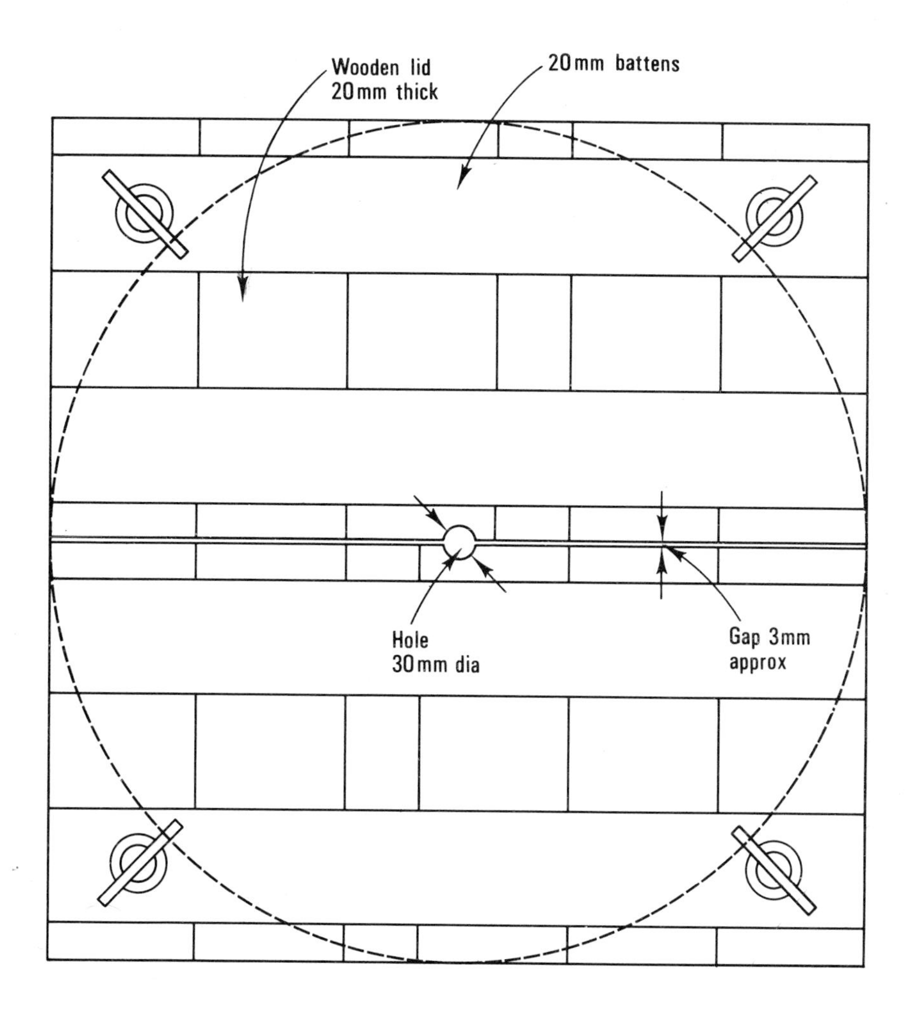

Figure 10.13 Shaft Plumbing - Drum and Lid

Appendix I

BIBLIOGRAPHY

(1) Published by H.M.S.O.

Constants, Formulae and Methods used in Transverse Mercator Projection.

(2) Obtainable from the Director General, Ordnance Survey,

Minor Control Point Albums
Bench Marks
Heights and Coordinates of Triangulation Stations.

(3) *Logarithms of Sines and Tangents for Every Second*

by Robert Shortrede, F.R.A.S.
Published by Vandyck Printers Limited.

(4) *Chambers Seven Figure Mathematical Tables.*

Appendix II

CONSTANTS

one second of arc	=	one sixteenth inch	in 1000 feet
	=	0.005 feet	in 1000 feet
	=	one inch	in three miles
one minute of arc	=	one inch	in 100 yards
one degree	=	one inch	in five feet

one second of arc = 5 mm in 1000 m
one minute of arc = 30 mm in 100 m
one degree = 20 mm in 1 m

Coefficient of expansion of steel = 0.0000062 per $^{\circ}$F
= 0.0000112 per $^{\circ}$C

Curvature of the earth's surface = 8 inches in one mile
= 12 cm in 1000 m

and is proportional to the square of the distance.

Offset to circular curve from a tangent = $L^2/2R$

where L = length of tangent and R = radius of circle.

Appendix III

CURVATURE AND REFRACTION CORRECTION FOR VERTICAL ANGLES

Horizontal distance (feet)	Correction (seconds)	Horizontal distance (feet)	Correction (seconds)
1000	4	6000	25
2000	8	7000	30
3000	13	8000	34
4000	17	9000	38
5000	21	10000	42

Horizontal distance (metres)	Correction (seconds)	Horizontal distance (metres)	Correction (seconds)
1000	14	6000	83
2000	28	7000	97
3000	42	8000	111
4000	56	9000	125
5000	69	10000	139

Appendix IV

CORRECTION – TO 1000 FEET LENGTH AT A GIVEN ALTITUDE

Altitude (feet)	Correction (feet)	Altitude (feet)	Correction (feet)
100	0.005	1100	0.052
200	0.010	1200	0.057
300	0.014	1300	0.062
400	0.019	1400	0.067
500	0.024	1500	0.071
600	0.029	1600	0.076
700	0.033	1700	0.081
800	0.038	1800	0.086
900	0.043	1900	0.090
1000	0.048	2000	0.095

CORRECTION – TO 1000 METRES LENGTH AT A GIVEN ALTITUDE

Altitude (metres)	Correction (metres)	Altitude (metres)	Correction (metres)
100	0.016	600	0.094
200	0.031	700	0.109
300	0.047	800	0.125
400	0.062	900	0.140
500	0.078	1000	0.156

Appendix V

PROPORTIONAL PARTS FOR SINGLE SECONDS OF ARC

Secs	2000	1000	900	800	700	600	500	400	300	Secs
1	33	17	15	13	12	10	8	7	5	59
2	67	33	30	27	23	20	17	13	10	58
3	100	50	45	40	35	30	25	20	15	57
4	133	67	60	53	47	40	33	27	20	56
5	167	83	75	67	58	50	42	33	25	55
6	200	100	90	80	70	60	50	40	30	54
7	233	117	105	93	82	70	58	47	35	53
8	267	133	120	107	93	80	67	53	40	52
9	300	150	135	120	105	90	75	60	45	51
10	333	167	150	133	117	100	83	67	50	50
11	367	183	165	147	128	110	82	73	55	49
12	400	200	180	160	140	120	100	80	60	48
13	433	217	195	173	152	130	108	87	65	47
14	467	233	210	187	163	140	117	93	70	46
15	500	250	225	200	175	150	125	100	75	45
16	533	267	240	213	187	160	133	107	80	44
17	567	283	255	227	198	170	142	113	85	43
18	600	300	270	240	210	180	150	120	90	42
19	633	317	285	253	222	190	158	127	95	41
20	667	333	300	267	233	200	167	133	100	40
21	700	350	315	280	245	210	175	140	105	39
22	733	367	330	293	257	220	183	147	110	38
23	767	383	345	307	268	230	192	153	115	37
24	800	400	360	320	280	240	200	160	120	36
25	833	417	375	333	292	250	208	167	125	35
26	867	433	390	347	303	260	217	173	130	34
27	900	450	405	360	315	270	225	180	135	33
28	933	467	420	373	327	280	233	187	140	32
29	967	483	435	387	338	290	242	193	145	31
30	1000	500	450	400	350	300	250	200	150	30
31	1033	517	465	413	362	310	258	207	155	29
32	1067	533	480	427	373	320	267	213	160	28
33	1100	550	495	440	385	330	275	220	165	27
34	1133	567	510	453	397	340	283	227	170	26
35	1167	583	525	467	408	350	292	233	175	25
36	1200	600	540	480	420	360	300	240	180	24
37	1233	617	555	493	432	370	308	247	185	23
38	1267	633	570	507	443	380	317	253	190	22
39	1300	650	585	520	455	390	325	260	195	21
40	1333	667	600	533	467	400	333	267	200	20
41	1367	683	615	547	478	410	342	273	205	19
42	1400	700	630	560	490	420	350	280	210	18
43	1433	717	645	573	502	430	358	287	215	17
44	1467	733	660	587	513	440	367	293	220	16
45	1500	750	675	600	525	450	375	300	225	15
46	1533	767	690	613	537	460	383	307	230	14
47	1567	783	705	627	548	470	392	313	235	13
48	1600	800	720	640	560	480	400	320	240	12
49	1633	817	735	653	572	490	408	327	245	11
50	1667	833	750	667	583	500	417	333	250	10
51	1700	850	765	680	595	510	425	340	255	9
52	1733	867	780	693	607	520	433	347	260	8
53	1767	883	795	707	618	530	442	353	265	7
54	1800	900	810	720	630	540	450	360	270	6
55	1833	917	825	733	642	550	458	367	275	5
56	1867	933	840	747	653	560	467	373	280	4
57	1900	950	855	760	665	570	475	380	285	3
58	1933	967	870	773	677	580	483	387	290	2
59	1967	983	885	787	688	590	492	393	295	1

Appendix VI

TACHEOMETRIC SURVEY TABLE

Vert. Angle	0'		10'		20'		30'		40'	
	H	V	H	V	H	V	H	V	H	V
0°	100.0	0.00	100.0	0.29	100.0	0.58	100.0	0.87	100.0	1.16
1	100.0	1.74	100.0	2.04	99.9	2.33	99.9	2.62	99.9	2.91
2	99.9	3.49	99.9	3.78	99.8	4.07	99.8	4.36	99.8	4.65
3	99.7	5.23	99.7	5.52	99.7	5.81	99.6	6.09	99.6	6.38
4	99.5	6.96	99.5	7.25	99.4	7.53	99.4	7.82	99.4	8.11
5	99.2	8.68	99.2	8.97	99.1	9.26	99.1	9.54	99.0	9.83
6	98.9	10.40	98.9	10.68	98.8	10.96	98.7	11.25	98.7	11.53
7	98.5	12.10	98.4	12.38	98.4	12.66	98.3	12.94	98.2	13.22
8	98.1	13.78	98.0	14.06	97.9	14.34	97.8	14.62	97.7	14.90
9	97.6	15.45	97.5	15.73	97.4	16.00	97.3	16.28	97.2	16.55
10	97.0	17.10	96.9	17.38	96.8	17.65	96.7	17.92	96.6	18.19
11	96.4	18.73	96.3	19.00	96.1	19.27	96.0	19.54	95.9	19.80
12	95.7	20.34	95.6	20.60	95.4	20.87	95.3	21.13	95.2	21.39
13	94.9	21.92	94.8	22.18	94.7	22.44	94.6	22.70	94.4	22.96
14	94.2	23.47	94.0	23.73	93.9	23.99	93.7	24.24	93.6	24.49
15	93.3	25.00	93.2	25.25	93.0	25.50	92.9	25.75	92.7	26.00
16	92.4	26.50	92.3	26.74	92.1	26.99	91.9	27.23	91.8	27.48
17	91.5	27.96	91.3	28.20	91.1	28.44	91.0	28.68	90.8	28.92
18	90.5	29.39	90.3	29.62	90.1	29.86	89.9	30.09	89.8	30.32
19	89.4	30.78	89.2	31.01	89.1	31.24	88.9	31.47	88.7	31.69
20	88.3	32.14	88.1	32.36	87.9	32.58	87.7	32.80	87.5	33.02
21	87.2	33.46	87.0	33.67	86.8	33.89	86.6	34.10	86.4	34.31
22	86.0	34.73	85.8	34.94	85.6	35.15	85.4	35.36	85.2	35.56
23	84.7	35.97	84.5	36.17	84.3	36.37	84.1	36.57	83.9	36.77
24	83.5	37.16	83.2	37.35	83.0	37.54	82.8	37.74	82.6	37.93
25	82.1	38.30	81.9	38.49	81.7	38.67	81.5	38.86	81.2	39.04

50'		Differences for Verticals									Vert.
H	V	1'	2'	3'	4'	5'	6'	7'	8'	9'	Angle
100.0	1.46	.03	.06	.09	.12	.15	.18	.21	.24	.27	0°
99.9	3.20	.03	.06	.09	.12	.15	.17	.20	.23	.26	1
99.8	4.94										2
99.6	6.67										3
99.3	8.40										4
99.0	10.11										5
98.6	11.81	.03	.06	.08	.11	.14	.17	.20	.22	.25	6
98.1	13.50										7
97.6	15.17										8
97.1	16.83										9
96.5	18.46										10
95.8	20.07	.03	.05	.08	.10	.13	.16	.18	.21	.24	11
95.1	21.66										12
94.3	23.22										13
93.5	24.75										14
92.6	26.25										15
91.6	27.72	.02	.05	.07	.09	.12	.14	.16	.18	.21	16
90.6	29.15										17
89.6	30.55										18
88.5	31.92										19
87.4	33.24										20
86.2	34.52	.02	.04	.06	.08	.10	.12	.14	.16	.18	21
84.9	35.76										22
83.7	36.96										23
82.4	38.11										24
81.0	39.22										25

Appendix VII

CRANDALL'S METHOD FOR CORRECTING LATITUDES AND DEPARTURES IN A TRAVERSE SO THAT BEARINGS OF LINES REMAIN UNCHANGED

In a side of length S and bearing A,

Let correction to length = cS
latitude = Y
departure = X

Then

Correction to latitude = cY
Correction to departure = cX

Assuming the weight of a linear measurement to be proportional to $1/S$, then by the method of least squares,

$$\sum(c^2S^2/S) = \sum(c^2S) = \text{a minimum}$$

Let $\sum(cY) = l$

$\sum(cX) = d$

Then

$$\sum(cS\delta c) = 0$$

and

$$\sum(Y\delta c) = 0$$

and

$$\sum(X\delta c) = 0$$

Multiplying the last two equations by $-f$ and $-g$ respectively, adding all three equations, and equating coefficients to zero:

$$c_1 = (fY_1 + gX_1)/S_1$$

$$c_2 = (fY_2 + gX_2)/S_2 \qquad \text{etc.}$$

Therefore

$$l = f\sum(Y^2/S) + g\sum(YX/S)$$

$$= f\sum(Y\cos A) + g\sum(S\cos A\sin A)$$

and

$$d = f\sum(YX/S) + g\sum(X^2/S)$$

$$= f\sum(S\cos A\sin A) + g\sum(X\sin A)$$

hence, f and g.

Then, corrections for latitude and departure are

$$l_1 = c_1Y_1 = (fY_1^2 + gY_1X_1)/S_1$$

$$= fY_1\cos A_1 + gS_1\cos A_1\sin A_1$$

$$d_1 = c_1X_1 = (fY_1X_1 + gX_1^2)/S_1$$

$$= fS_1\cos A_1\sin A_1 + gX_1\sin A_1$$

Appendix VIII

NOTES ON THE USE OF THE GEODIMETER (AGA MODEL 4B)

(A) PROCEDURE BEFORE INSTRUMENT CHECKS AND MEASURING

(*a*) Leave front cover closed
(*b*) Allow to warm up, (wait until top LH panel light goes out)
Note: wait thrice this length of time before measuring.
(*c*) Light OFF; F 1; PH 1; ADJ; CAL; LIGHT SENS out.
(*d*) INST 1 volts 6.3.
(*e*) INST 2 milliamps 30 to 40.
(*f*) INST 3 Zero nullmeter (zero ADJ).
(*g*) Switch to PH 2 check nullmeter zeroing. It should not be more than half way to its maximum deflection on the scale.
(*h*) If not satisfactory, adjust R.69 (inside, top LHS).

Note: It is <u>most</u> unlikely that the nullmeter will be unsatisfactory at (*g*).

(B) PROCEDURE FOR SIGHTING TARGETS

(*a*) Open lower front cover.
(*b*) Aperture (Front RHS) Bright, Small; Dull; Medium; Night, Large.
(*c*) Light ON; F 1; Ph 1; ADJ; REFL; INST 3; Telescope out.
(*d*) Peepsight own or reflector attendants torch.
(*e*) Focus telescope, align geodimeter (reflector in centre of Kerr Cell) RECORD FOCUS DISTANCE.
(*f*) MEAS. LIGHT SENS 0.8 to 1.0.
(*g*) Align to obtain minimum on instrument meter. Repeat (*f*) and (*g*) alternately for accurate alignment.

(C) PROCEDURE FOR MEASURING

(*a*) FIRST, record ALL the data at the head of the computation sheet.
(*b*) INST. 2. Tune Kerr Cell (+/- switch, front centre) to obtain minimum on instrument meter.
(*c*) INST. 3. CAL; LIGHT SENS 0.8 to 1.0.
(*d*) Record Null readings Fl; Ph 1. 2. 3. 4. (Note: Book sign s or o for Ph 1 <u>and</u> Ph 3). NEVER move the phase switch anticlockwise.
(*e*) REFL; LIGHT SENS 0.8 to 1.0.
(*f*) Record Null readings F 1; Ph 1. 2. 3. 4.
(*g*) Repeat (*b*) to (*e*), F2 and F3.
(*h*) BEFORE packing up, CHECK all the data at the head of the computation sheet.

MEASURING - SUMMARY

MEAS.					
"	F 1		INST 2		Kerr Cell
"	"	CAL	INST 3		Light Sens
"	"	"	"	Ph 1-4	Read Null
"	"	REFL	"		Light Sens
"	"	"	"	Ph 1-4	Read Null
"	F 2		INST 2		Kerr Cell
"	"	REFL	INST 3		Light Sens
"	"	"	"	Ph 1-4	Read Null
"	"	CAL	"		Light Sens
"	"	"	"	Ph 1-4	Read Null
"	F 3		INST 2		Kerr Cell
"	"	CAL	INST 3		Light Sens
"	"	"	"	Ph 1-4	Read Null
"	"	REFL	"		Light Sens
"	"	"	"	Ph 1-4	Read Null

(D) PROCEDURE FOR SETTING UP REFLECTORS

(*a*) *Reflectors Used*	*Slope Distance up to* by day	by night
Plastic Reflector	150 m	1000 m
One Prism	300	3000
Three Prisms	700 to 1000	7000
Seven Prisms	1000 to 2000	10000
(*b*) *Wedges Used*	*Slope Distances* from	to
400"	100 m	200 m
200"	200	400
100"	400	800
50"	800	1600
None	1600	Upwards

(*c*) Prisms must be mounted in their holders with dots upwards. Wedges must be placed with straight edge vertical.

(*d*) For vertical angles up to 5°, set reflector box upright and plumb over station. Height of reflector is measured to axis of circular box.

For steeper slopes, tilt the box so that its axis points to the geodimeter. The point where the axis of the base screw meets the axis of the box must be vertically above the station. Measure to this point to obtain the height of reflector.

(*e*) Where the background is very light, put a black screen 500 mm square behind the reflector. This is only necessary in extreme cases.

Note: Section (*a*) is a guide only. Atmospheric conditions vary the effectiveness of the transmitted light and consequently the range.

(E) PROCEDURE ON REPLACING PROJECTOR LAMP

(*a*) Take off front cover.
(*b*) Screw lamp in tightly.
(*c*) Turn the holder so that the filament is not obscured by its own supports.
(*d*) Do not replace lampguard upside down (see hole).
(*e*) Light ON; REFL; MEAS.
(*f*) Look into transmitter (LH Tube) rotate polaroid to make light yellow.
(*g*) Move Lampholder along its axis to align with Kerr Cell.
Move Lampholder to or from you to align with Kerr Cell.
(*h*) Reset Polaroid to remove all yellowness from light.
(*i*) Replace front cover.

Note: Field glasses may be used for (*f*) and (*g*).

(F) PROCEDURE FOR INTERNAL OPTICAL ALIGNMENT

(*a*) Set a good point light, dark background, 1000' away.
(*b*) Remove front cover.
(*c*) Switch to small aperture (Front RHS).
(*d*) INST 3; REFL; ADJ.
(*e*) Peepsight far light.
(*f*) Pull out telescope, FOCUS far Light.
(*g*) LIGHT SENS to give 0.8 to 1.0 on instrument meter.
(*h*) Align geodimeter to give minimum reading on instrument meter repeat (*g*) and (*h*) alternately for accurate alignment.
(*i*) Light ON. Look through telescope, and if necessary move Kerr Cell on to far light (Kerr Cell knob, front centre).
(*j*) MEAS.
(*k*) Look into transmitter, rotate polaroid to make light yellow.
(*l*) Move lampholder along its axis to align with Kerr Cell.
Move lampholder to or from you to align with Kerr Cell.
(*m*) Reset Polaroid to remove all yellowness from light.
(*n*) Replace front cover.

Note: Field glasses may be used for (*k*) and (*l*).

(G) PROCEDURE FOR MAINTENANCE

(1) General physical check.
(*a*) Screw on caps must be lightly greased after cleaning.
(*b*) Dessicators. Geodimeter Unit, one inside top RHS.
Power Unit, one on top.
This gel should be blue; if not, heat at about 95°C for about 30 minutes.
(*c*) Lights on Geodimeter. One panel light; back, top LHS.
One panel light; back, top RHS.
One peep sight light; inside bottom LHS.
Change when necessary.
(*d*) Valves, look for cracked or blackened tubes (see procedure H).

(H) PROCEDURE FOR FAULT FINDING - 1. GENERAL

The geodimeter gives consistent readings everywhere with only small variations when it is in good order and adjustment. There should be no gradual falling off in sensitivity; therefore faults will be indicated by immediate and large variations from the normal readings.

Provided readings are obtainable at all, even with a spread, the selection of Frequencies and Phase positions is such that a wrong length will not result and its accuracy will be reduced by only a small amount. Faults which could produce a wrong length cut out the geodimeter response so that a set of readings cannot be obtained.

The variations are caused more often by bad alignment or adjustment than by electrical, etc., faults.

Therefore:

(1) Check for flat battery.
(2) Re-check procedure (A) and (F) carefully.
(3) Only then attempt fault finding below.
(4) If this fails:
    (1) Do not alter further any, setting, adjustments, etc., at all.
    (2) Retain all suspect parts.
    (3) Have ready full notes of all actions taken, and the latest computation sheets.
    (4) Inform the manufacturer.

(H) PROCEDURE FOR FAULT FINDING - 2. NORMAL READINGS

INST 1 6.3 volts on instrument meter.
INST 2 30 to 40 milliamps on instrument meter.
INST 3 Maximum 1.2 on instrument meter when receiving strong light.
INST 3 Instrument meter will vary from about zero to 1.2 as any incoming light is varied, or as LIGHT SENS is varied.
INST 3 Null meter will dip one way or the other on turning the delay dial when modulated light is being received.
INST 3 CAL. Delay Dial readings always give the same sign and they are constant (to about ± 5 divisions) throughout the life of a given geodimeter.
INST 3 CAL. or REFL. Groups of Delay Dial should not spread by more than about 10 divisions at each frequency.

(H) PROCEDURE FOR FAULT FINDING - 3. FAULTY READINGS

(*a*) FAULT INST 1 and 2 satisfactory, but at INST 3 instrument meter much higher than 1.2 and can be varied only a little.
REMEDY Change Valve V7.

(*b*) FAULT INST 3, MEAS, REFL, Light emission satisfactory, reflex seen by eye, but no reading on instrument meter.
REMEDY (1) Check Kerr Cell position (See Procedure F).
or (2) Check Phototube by waving torch in front of receiving tube (RHS). Instrument meter should vary considerably. If not inform manufacturer.

(*c*) FAULT INST 3, MEAS. CAL. Item (*b*) satisfactory, but little or no response on instrument meter.
REMEDY Clean. Straighten or adjust stainless steel mirror so that transmitter light is reflected round into the receiving tube to give a strong response on the instrument meter.

(*d*) FAULT Measuring CAL; delay Dial readings all lower than normal and sign may have changed; INST 2. Instrument meter very high (probably off the dial).
REMEDY (1) Clean Phase switches by rotating clockwise several times.
or (2) Check Polaroid position (See Procedure F) (when wrong, delay Dial readings lower, sign changes).
or (3) Change Valves V16 and V17.

(*e*) FAULT Measuring CAL. Delay Dial readings all lower than normal and sign may have changed; INST 2 Instrument meter very low.
REMEDY (1) As Item (*d*)(1).
or (2) As Item (*d*)(2).
or (3) Change Valve V19.

(*f*) FAULT Measure REFL. or CAL. nullmeter will not move in two of the phases.
REMEDY Change Valves V8 and V9.

(*g*) FAULT Measure REFL. or CAL. nullmeter will not move in any of the phases.
REMEDY (1) Change Valve V15.
or (2) Change Valves V8 and V9.

Note: Valves V8 and V9 are a matched pair, i.e. they must always be removed as a pair and replaced by the spare pair which are taped together. Should one of the removed pair be sound it may be retained as a spare for valves V7, V15, V16, which are the same type.

(*h*) FAULT Large spread between the two pairs of Delay Dial readings on each frequency.
REMEDY (1) Clean Phase switches by rotating clockwise several times.
or (2) Check nullmeter ZERO ADJ (See Procedure A).
or (3) Diodes may need check by the manufacturer.

Final note

Do not touch resistors, inductances, condensers, diodes, valves, photo-tube or any items not mentioned in these instructions.

Appendix IX

AGA

DISTANCE MEASURING

Geodimeter Model 6 serial No. 665555 Time --- Date ----

Corrections		
Geod. eccentr.	NIL	m
Reflector eccentr.	NIL	m
Geod. constant	− 0·130	m
Reflector constant	− 0·030	m
Atm. corr.	+ 0·010	m
	+ 0·010	m
	− 0·160	m
Sum of corr.	− 0·150	m

Temperature 10 °C

Bar. press 761 mm Hg

Atm. corr. 15 10^{-6}D

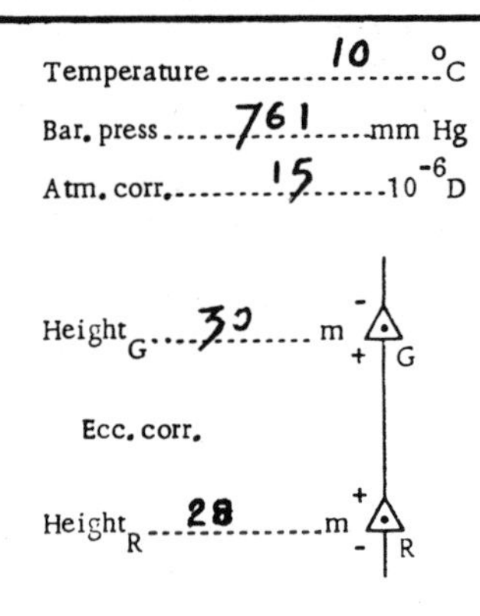

Height$_G$ 30 m

Ecc. corr.

Height$_R$ 28 m

Geodimeter station Offord

Reflector station Cluny

Observer ACK

Recorder RHQC

Cal. table date 19·3·71

Approx. dist. 700 m

Area Burnley

Type of refl. 3 prisms

Visibility: Good Fair Poor

Phase	U1 s/o	C_1	s/o	R_1	U2 s/o	C_2	s/o	R_2	U3 s/o	C_3	s/o	R_3
1	S	257	S	116	S	210	S	445	S	261	S	709
2		256		115		208		446		261		708
3	S	271	S	122	S	220	S	459	S	273	S	720
4		272		122		221		456		273		719
Sum of phases 2 and 3		527		237		428		905		534		1428
Sum of phases 1 and 4		529		238		431		901		534		1428
Sum of phases 1,2,3,4		1056		475		859		1806		1068		2856

	U_1			U_2			U_3		
Meters from table	S 1·324	S	0·589	S 1·104	S	2·040	S 0·980	S	3·085
If C>R add U and change "s" to "o" or "o" to "s"	$(+U_1)$		(+2,500)	$(+U_2)$		~~(+2,494)~~	$(+U_3)$		~~(+2,381)~~
	$R_1+(U_1)$	o	3·089	$R_2+(U_2)$	S	2·040	$R_3+(U_3)$	S	3·085
	$-C_1$		−1·324	$-C_2$		−1·104	$-C_3$		−·980
	$R_1-C_1(+U_1)$	s/o	1·765	$R_2-C_2(+U_2)$	s/s	0·936	$R_3-C_3(+U_3)$	s/s	2·105
If "so" or "os" add U	$(+U_1)$		(+2,500)	$(+U_2)$		~~(+2,494)~~	$(+U_3)$		~~(+2,381)~~
(L) = R−C (+U) (+U)			4·265	(L_2)		0·936	(L_3)		2·105
If $(L_2)<L_1$, $(L_3)<L_1$ add. $2U_2$ resp. $2U_3$				$(+2U_2)$		(+4,988)	$(+2U_3)$		(+4,762)
		L_1 =	4·265		L_2 =	5·924		L_3 =	6·867
					$-L_1$ = −	4·265		$-L_1$ = −	4·265

$L_1+L_{2k}+L_{3k}$	L_{2k}	L_{3k}		
L_1 = 4·265	L_2 = 5·924	L_3 = 6·867	$A=L_2-L_1$ = 1·659	$B=L_3-L_1$ = 2·602
L_{2k} = 4·291	$-K_2$ = 1·633	$-K_3$ = 2·619	400 A = 663·6	+20 B 52·04
L_{3k} = 4·248	L_{2k} = 4·291	L_{3k} = 4·248	−F 55	(F) = 21 B = 54·64
ΣL = 12·804	ΣL : 3 = 4·268		(E) = 400 A − F = 608·6	F (nearest multiple of 5) = 55
Formulas (see table)	D' = E + F = 655·000		E (nearest hundreds) = 600	

Formulas	
$D' = E + F$	n x 2000 = —
$K_2 = D' \cdot 0{,}002493766$	Σ = 659·268
$K_3 = F \cdot 0{,}0476190$	Sum of corr. = −0·150
$L_{2k} = L_2 - K_2$ $L_{3k} = L_3 - K_3$	L + D' + n · 2000 + + corr. = 659·118

Remarks:

Appendix X

GEODIMETER MODEL 6 CALIBRATION TABLES

(1) DATE 710319 NUMBER 665555 CONSTANT - .130 METRE FREQ 1

	00	10	20	30	40	50	60	70	80	90
0	0.000	.011	.022	.034	.045	.056	.068	.079	.091	.102
1	.114	.126	.137	.149	.161	.173	.185	.197	.209	.221
2	.233	.245	.258	.270	.283	.295	.308	.321	.334	.346
3	.359	.372	.385	.397	.410	.423	.436	.449	.462	.475
4	.488	.501	.515	.528	.541	.555	.569	.583	.596	.610
5	.623	.636	.650	.663	.676	.689	.702	.715	.728	.742
6	.755	.768	.782	.795	.808	.822	.836	.849	.862	.876
7	.889	.902	.915	.929	.942	.955	.968	.981	.994	1.007
8	1.020	1.033	1.045	1.058	1.071	1.083	1.096	1.109	1.121	1.133
9	1.145	1.157	1.169	1.181	1.192	1.204	1.215	1.227	1.238	1.250
10	1.261	1.272	1.284	1.295	1.306	1.317	1.328	1.339	1.350	1.361
11	1.372	1.383	1.393	1.404	1.415	1.425	1.436	1.446	1.456	1.467
12	1.477	1.487	1.497	1.508	1.518	1.528	1.538	1.547	1.557	1.567
13	1.577	1.587	1.597	1.606	1.616	1.626	1.636	1.645	1.655	1.665
14	1.674	1.684	1.693	1.703	1.712	1.721	1.731	1.740	1.749	1.759
15	1.768	1.777	1.787	1.796	1.806	1.815	1.824	1.834	1.843	1.853
16	1.862	1.871	1.881	1.890	1.900	1.909	1.919	1.928	1.938	1.948
17	1.957	1.967	1.976	1.986	1.995	2.005	2.014	2.024	2.034	2.043
18	2.053	2.063	2.073	2.082	2.092	2.102	2.112	2.122	2.132	2.142
19	2.152	2.162	2.172	2.182	2.193	2.203	2.213	2.224	2.234	2.245
20	2.255	2.266	2.276	2.287	2.298	2.309	2.320	2.331	2.342	2.353
21	2.364	2.375	2.387	2.398	2.410	2.421	2.433	2.444	2.456	2.468
22	2.480	2.492	2.504	2.516	2.529	2.541	2.554	2.567	2.579	2.592
23	2.604	2.617	2.629	2.642	2.654	2.667	2.680	2.693	2.706	2.719
24	2.731	2.743	2.756	2.768	2.781	2.793	2.805	2.817	2.830	2.842
25	2.854	2.866	2.878	2.891	2.903	2.915	2.927	2.939	2.952	2.964
26	2.976	2.988	3.001	3.013	3.025	3.038	3.050	3.063	3.075	3.088
27	3.100	3.113	3.125	3.138	3.150	3.163	3.176	3.189	3.202	3.214
28	3.227	3.240	3.253	3.265	3.278	3.291	3.305	3.318	3.331	3.343
29	3.356	3.369	3.381	3.394	3.407	3.419	3.432	3.445	3.457	3.470
30	3.482									

(2) DATE 710319 NUMBER 665555 CONSTANT - .130 METRE FREQ 2

	00	10	20	30	40	50	60	70	80	90
0	0.000	.012	.024	.036	.048	.060	.072	.085	.097	.109
1	.122	.135	.147	.160	.173	.186	.199	.212	.225	.238
2	.251	.264	.277	.290	.303	.316	.329	.342	.356	.369
3	.382	.395	.408	.422	.435	.448	.462	.475	.488	.502
4	.515	.528	.542	.555	.568	.581	.595	.608	.621	.635
5	.648	.661	.675	.688	.702	.716	.729	.743	.757	.770
6	.783	.796	.809	.822	.835	.848	.861	.873	.886	.898
7	.911	.923	.936	.948	.961	.973	.985	.997	1.009	1.021
8	1.033	1.045	1.057	1.069	1.081	1.093	1.105	1.116	1.128	1.140
9	1.152	1.164	1.176	1.188	1.200	1.212	1.224	1.236	1.248	1.260
10	1.272	1.284	1.296	1.308	1.320	1.332	1.344	1.357	1.368	1.380
11	1.391	1.402	1.413	1.424	1.435	1.446	1.457	1.467	1.478	1.488
12	1.498	1.508	1.518	1.528	1.537	1.546	1.556	1.565	1.574	1.583
13	1.592	1.601	1.609	1.618	1.627	1.634	1.642	1.650	1.658	1.667
14	1.675	1.683	1.692	1.700	1.708	1.717	1.725	1.734	1.742	1.751
15	1.759	1.768	1.776	1.785	1.793	1.802	1.810	1.819	1.827	1.836
16	1.845	1.854	1.863	1.872	1.881	1.890	1.899	1.908	1.917	1.927
17	1.936	1.945	1.955	1.965	1.974	1.984	1.994	2.004	2.014	2.024
18	2.034	2.044	2.055	2.066	2.076	2.087	2.099	2.110	2.121	2.132
19	2.143	2.154=	2.165	2.177	2.188	2.200	2.211	2.223	2.234	2.246
20	2.257	2.268	2.280	2.291	2.303	2.314	2.325	2.336	2.348	2.360
21	2.371	2.383	2.394	2.406	2.418	2.429	2.441	2.453	2.465	2.477
22	2.489	2.501	2.514	2.526	2.539	2.552	2.565	2.579	2.591	2.604
23	2.617	2.630	2.643	2.655	2.668	2.681	2.694	2.707	2.720	2.733
24	2.745	2.757	2.770	2.782	2.795	2.806	2.818	2.830	2.843	2.855
25	2.867	2.879	2.892	2.904	2.916	2.929	2.941	2.954	2.966	2.979
26	2.991	3.004	3.016	3.029	3.041	3.054	3.067	3.080	3.093	3.105
27	3.118	3.131	3.144	3.156	3.169	3.182	3.195	3.208	3.221	3.234
28	3.247	3.260	3.273	3.286	3.299	3.312	3.325	3.339	3.351	3.364
29	3.377	3.390	3.402	3.415	3.428	3.440	3.453	3.465	3.477	3.490
30	3.502									

(3) DATE 710319 NUMBER 665555 CONSTANT -.130 METER FREQ 3

	00	10	20	30	40	50	60	70	80	90
0	0.000	.009	.019	.028	.037	.047	.056	.065	.075	.084
1	.093	.102	.111	.121	.130	.139	.148	.157	.166	.175
2	.184	.193	.202	.211	.220	.229	.238	.247	.256	.265
3	.274	.283	.292	.301	.309	.318	.327	.336	.345	.353
4	.362	.371	.379	.388	.397	.405	.414	.422	.431	.439
5	.448	.456	.465	.473	.482	.490	.498	.507	.515	.524
6	.532	.540	.549	.557	.566	.574	.583	.592	.600	.609
7	.617	.626	.634	.643	.651	.659	.667	.675	.684	.693
8	.702	.711	.720	.729	.739	.748	.758	.767	.777	.787
9	.797	.807	.817	.828	.838	.849	.860	.870	.881	.892
10	.903	.914	.925	.936	.948	.959	.971	.982	.994	1.005
11	1.017	1.029	1.040	1.052	1.064	1.076	1.087	1.099	1.111	1.124
12	1.136	1.149	1.161	1.174	1.187	1.200	1.213	1.226	1.239	1.253
13	1.267	1.281	1.296	1.310	1.325	1.341	1.357	1.373	1.388	1.403
14	1.418	1.433	1.448	1.463	1.479	1.494	1.510	1.526	1.541	1.555
15	1.570	1.585	1.599	1.613	1.628	1.641	1.655	1.668	1.683	1.697
16	1.711	1.725	1.740	1.754	1.768	1.783	1.798	1.812	1.826	1.841
17	1.855	1.869	1.884	1.898	1.912	1.926	1.940	1.954	1.968	1.983
18	1.997	2.011	2.026	2.040	2.055	2.070	2.086	2.101	2.115	2.129
19	2.143	2.157	2.170	2.183	2.196	2.209	2.221	2.234	2.246	2.259
20	2.271	2.283	2.295	2.307	2.319	2.331	2.342	2.354	2.365	2.377
21	2.388	2.399	2.411	2.422	2.433	2.445	2.456	2.467	2.478	2.489
22	2.500	2.511	2.522	2.532	2.543	2.553	2.564	2.574	2.585	2.595
23	2.605	2.615	2.625	2.635	2.645	2.654	2.664	2.674	2.683	2.693
24	2.702	2.711	2.721	2.730	2.739	2.748	2.757	2.767	2.775	2.784
25	2.793	2.802	2.810	2.819	2.827	2.836	2,844	2.852	2.861	2.869
26	2.877	2.885	2.893	2.901	2.909	2.917	2.925	2.933	2.941	2.949
27	2.957	2.965	2.973	2.981	2.990	2.998	3.006	3.014	3.023	3.031
28	3.039	3.047	3.055	3.064	3.072	3.080	3.089	3.097	3.105	3.113
29	3.121	3.129	3.137	3.145	3.153	3.161	3.168	3.176	3.184	3.191
30	3.199									

(4) $K_2 = D \cdot 0.002493766$ for every 5 m up to 2000 m

×100 / ×1	0	1	2	3	4	5	6	7	8	9
00	0.000	0.249	0.499	0.748	0.998	1.247	1.496	1.746	1.995	2.244
05	0.012	0.262	0.511	0.761	1.010	1.259	1.509	1.758	2.007	2.257
10	0.025	0.274	0.524	0.773	1.022	1.272	1.521	1.771	2.020	2.269
15	0.037	0.287	0.536	0.786	1.035	1.284	1.534	1.783	2.032	2.282
20	0.050	0.299	0.549	0.798	1.047	1.297	1.546	1.796	2.045	2.294
25	0.062	0.312	0.561	0.810	1.060	1.309	1.559	1.808	2.057	2.307
30	0.075	0.324	0.574	0.823	1.072	1.322	1.571	1.820	2.070	2.319
35	0.087	0.337	0.586	0.835	1.085	1.334	1.584	1.833	2.082	2.332
40	0.100	0.349	0.599	0.848	1.097	1.347	1.596	1.845	2.095	2.344
45	0.112	0.362	0.611	0.860	1.110	1.359	1.608	1.858	2.107	2.357
50	0.125	0.374	0.623	0.873	1.122	1.372	1.621	1.870	2.120	2.369
55	0.137	0.387	0.636	0.885	1.135	1.384	1.633	1.883	2.132	2.382
60	0.150	0.399	0.648	0.898	1.147	1.397	1.646	1.895	2.145	2.394
65	0.162	0.411	0.661	0.910	1.160	1.409	1.658	1.908	2.157	2.406
70	0.175	0.424	0.673	0.923	1.172	1.421	1.671	1.920	2.170	2.419
75	0.187	0.436	0.686	0.935	1.185	1.434	1.683	1.933	2.182	2.431
80	0.200	0.449	0.698	0.948	1.197	1.446	1.696	1.945	2.195	2.444
85	0.212	0.461	0.711	0.960	1.209	1.459	1.708	1.958	2.207	2.456
90	0.224	0.474	0.723	0.973	1.222	1.471	1.721	1.970	2.219	2.469
95	0.237	0.486	0.736	0.985	1.234	1.484	1.733	1.983	2.232	2.481
100										

10	11	12	13	14	15	16	17	18	19
2.494	2.743	2.993	3.242	3.491	3.741	3.990	4.239	4.489	4.738
2.506	2.756	3.005	3.254	3.504	3.753	4.002	4.252	4.501	4.751
2.519	2.768	3.017	3.267	3.516	3.766	4.015	4.264	4.514	4.763
2.531	2.781	3.030	3.279	3.529	3.778	4.027	4.277	4.526	4.776
2.544	2.793	3.042	3.292	3.541	3.791	4.040	4.289	4.539	4.788
2.556	2.805	3.055	3.304	3.554	3.803	4.052	4.302	4.551	4.800
2.569	2.818	3.067	3.317	3.566	3.815	4.065	4.314	4.564	4.813
2.581	2.830	3.080	3.329	3.579	3.828	4.077	4.327	4.576	4.825
2.594	2.843	3.092	3.342	3.591	3.840	4.090	4.339	4.589	4.838
2.606	2.855	3.105	3.354	3.603	3.853	4.102	4.352	4.601	4.850
2.618	2.868	3.117	3.367	3.616	3.865	4.115	4.364	4.613	4.863
2.631	2.880	3.130	3.379	3.628	3.878	4.127	4.377	4.626	4.875
2.643	2.893	3.142	3.392	3.641	3.890	4.140	4.389	4.638	4.888
2.656	2.905	3.155	3.404	3.653	3.903	4.152	4.401	4.651	4.900
2.668	2.918	3.167	3.416	3.666	3.915	4.165	4.414	4.663	4.913
2.681	2.930	3.180	3.429	3.678	3.928	4.177	4.426	4.676	4.925
2.693	2.943	3.192	3.441	3.691	3.940	4.190	4.439	4.688	4.938
2.706	2.955	3.204	3.454	3.703	3.953	4.202	4.451	4.701	4.950
2.718	2.968	3.217	3.466	3.716	3.965	4.214	4.464	4.713	4.963
2.731	2.980	3.229	3.479	3.728	3.978	4.227	4.476	4.726	4.975
									4.988

K_3 = F 0.0476190 for every 5 m up to 100 m

m	K_3
00	0.000
05	0.238
10	0.476
15	0.714
20	0.952
25	1.190
30	1.429
35	1.667
40	1.905
45	2.143
50	2.381
55	2.619
60	2.857
65	3.095
70	3.333
75	3.571
80	3.810
85	4.048
90	4.286
95	4.524
100	4.762

Appendix XI

GEODIMETER MODEL 6
NOMOGRAM FOR THE ATMOSPHERIC CORRECTION

°C °F | parts x 10^{-6} | inches Hg mm Hg

°C: +40, +35, +30, +25, +20, +15, +10, +5, ±0, −5, −10, −15, −20, −25, −30, −35, −40

°F: +100, +95, +90, +85, +80, +75, +70, +65, +60, +55, +50, +45, +40, +35, +30, +25, +20, +15, +10, +5, ±0, −5, −10, −15, −20, −25, −30, −35, −40

parts x 10^{-6}: +150, +140, +130, +120, +110, +100, +90, +80, +70, +60, +50, +40, +30, +20, +10, ±0, −10, −20, −30, −40, −50, −60, −70

inches Hg: 18, 19, 20, 21, 22, 23, 24, 25, 26, 27, 28, 29, 30, 31, 32

mm Hg: 450, 460, 470, 480, 490, 500, 510, 520, 530, 540, 550, 560, 570, 580, 590, 600, 610, 620, 630, 640, 650, 660, 670, 680, 690, 700, 710, 720, 730, 740, 750, 760, 770, 780, 790, 800, 810, 820

Place a rule so that it intersects the temperature and pressure scales at the temperature and pressure values recorded. The correction in parts per million is read off on the center scale. For a barometer graduated in millibars, multiply by 0.75 and use the mm Hg scale.

Appendix XII

THE KERN ME 3000 MEKOMETER

Daylight measuring range:	3000 m
Accuracy (m.s.e.)	± (1 mm + 1 ppm)
Country of origin	Switzerland (NRDC licence)
Carrier wavelength	White
Measuring frequencies:	500 MHz
Diameter of transmitting objective	40 mm
Diameter of receiving objectives	40 mm
Focal length:	140 mm
Operation:	Manually operated phase meter
Distance display:	Numerical counter display
Measuring time:	2 minutes
Power consumption with 12 V:	25 W
Weight	17 kg
Temperature range:	- 20 to + 40°C

Appendix XIII

AGA

DISTANCE MEASURING

Geodimeter Model 6A serial No. .... Time .... Date ....

Corrections

Geod. eccentr. .... m

Reflector eccentr. .... m

Geod. constant .... m

Reflector constant .... m

Atm. corr. .... m

Sum + ... m

Sum - .... m

Sum of corr. .... m

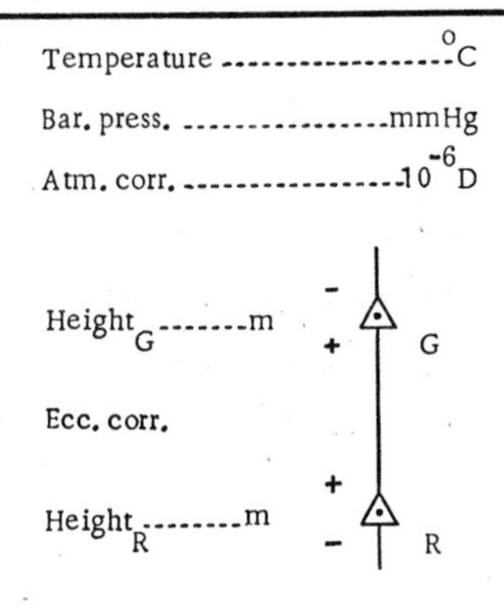

Geodimeter station ....

Reflector station ....

Observer ....

Recorder ....

Approx. dist. ....

Area ....

Type of refl. ....

Visibility: Good Fair Poor

Phase	U_1				U_2				U_3			
	s/o	C_1	s/o	R_1	s/o	C_2	s/o	R_2	s/o	C_3	s/o	R_3
1												
2												
3												
4												
Sum of phases 2 and 3												
Sum of phases 1 and 4												
Sum of phases 1,2,3,4												

	U_1				U_2				U_3			
Sum of phases 1,2,3,4 : 1000												
If C>R add. U and change "s" to "o" or "o" to "s"	$(+U_1)$			(+2,500)	$(+U_1)$			(+2,500)	$(+U_1)$			(+2,500)
	$R_1(+U_1)$				$R_2(+U_1)$				$R_3(+U_1)$			
	$-C_1$				$-C_2$				$-C_3$			
	$R_1-C_1(+U_1)$				$R_2-C_2(+U_1)$				$R_3-C_3(+U_1)$			
If "so" or "os" add U_1	$(+U_1)$			(+2,500)	$(+U_1)$			(+2,500)	$(+U_1)$			(+2,500)
$L' = R-C(+U_1)(+U_1)$					L_2'				L_3'			
					$-L_2'$: 400			−	$-L_3'$: 20			−
					L_2''							
									$+L_3'$: (20×20)			+
									L_3''			
If $L_2''<L_1$, $L_3''<L_1$ add $2U_2$ resp. $2U_3$					$(+2U_2)$			(+4,988)	$(+2U_3)$			(+4,762)
			L_1 =			L_2 =				L_3 =		
						$-L_1$ =				$-L_1$ =		

$L_1 + L_{2k} + L_{3k}$	L_{2k}	L_{3k}
L_1 =	L_2 =	L_3 =
L_{2k} =	$-K_2$ = −	$-K_3$ = −
L_{3k} =	L_{2k} =	L_{3k} =

A = $L_2 - L_1$ =	B = $L_3 - L_1$ =
400 A =	+ 20B =
− F	(F) = 21B =
(E) = 400 A − F =	F (nearest multiple of 5) =
E (nearest hundreds)=	

ΣL =	ΣL : 3 =
Formulas (see table)	D = E + F =
$D = E + F$ $P = (n \times 2000)$ $K_2 = D \times 0{,}002493766$ $K_3 = F \times 0{,}0476190$ $L_{2k} = L_2 - K_2$ $L_{3k} = L_3 - K_3$	P (n × 2000) = Σ = Sum of corr. = L + D + P + corr. =

Remarks: ....